The … stry and the European Community

The Construction Industry and the European Community

N. F. Spencer Chapman
and
C. Grandjean

OXFORD
BSP PROFESSIONAL BOOKS
LONDON EDINBURGH BOSTON
MELBOURNE PARIS BERLIN VIENNA

BSP Professional Books
A division of Blackwell Scientific
Publications Ltd
Editorial offices:
Osney Mead, Oxford OX2 0EL
25 John Street, London WC1N 2BL
23 Ainslie Place, Edinburgh EH3 6AJ
3 Cambridge Center, Cambridge,
MA 02142, USA
54 University Street, Carlton,
Victoria 3053, Australia

First published 1991

Set by DP Photosetting, Aylesbury, Bucks
Printed and bound in Great Britain by
Hartnolls Ltd, Bodmin, Cornwall

DISTRIBUTORS

Marston Book Services Ltd
PO Box 87
Oxford OX2 0DT
(*Orders:* Tel: 0865 791155
Fax: 0865 791927
Telex: 837515)

USA
Blackwell Scientific Publications, Inc.
3 Cambridge Center
Cambridge, MA 02142
(*Orders:* Tel: (800) 759-6102)

Canada
Oxford University Press
70 Wynford Drive
Don Mills
Ontario M3C 1J9
(*Orders:* Tel: (416) 441-2941)

Australia
Blackwell Scientific Publications
(Australia) Pty Ltd
54 University Street
Carlton, Victoria 3053
(*Orders:* Tel: (03) 347-0300)

British Library
Cataloguing in Publication Data

Chapman, N. F. Spencer
The construction industry and the
European Community.
I. Title II. Grandjean, C.
338.47624094

ISBN 0-632-03121-2

Contents

Preface

A welter of information is being published by and about the European Community, its Member States and industries. Despite that, people have considerable difficulty in obtaining exactly the information they need. It is often costly and not available in most libraries; and the widespread use of jargon and unfamiliar concepts adds to the problem.

This book has therefore been written to provide a single, readable guide for those in the construction industry who want to understand what is happening, who want to form their own views about how to respond and who want to be able to find further information. It is intended primarily for contractors, consultants and clients as well as product manufacturers, suppliers and sub-contractors. It should also be valuable for students of civil engineering, architecture, building and related disciplines, particularly those studying specialist options or wanting to take advantage of the opportunities available in the emerging Euro construction businesses. Plain English is used and jargon has been avoided as much as possible. The whole subject is riddled with abbreviations and acronyms but a key to these has been given on page xii.

The book aims for a European perspective although the UK is often used for comparison, and the section on information sources describes UK as well as EC sources. The authors come from different Member States, have dissimilar cultural backgrounds and both benefit from diverse and international experience. Their efforts could well have resulted in two books but they hope that the combined result is more comprehensive, balanced and interesting. The book is substantially factual but views have been expressed following interviews with a number of senior people in the industry or connected with it and through lively debate between the authors.

Three early chapters describe the European Community, its institutions and the often fascinating developments in Community law which, in conjunction with the '1992' single market programme, are having far-reaching effects on Community residents and businesses. One chapter provides an overview of the Community construction industry and comparisons amongst national systems, economies and industries. A series of studies on each of the twelve Member States examines the economic, cultural, social and legal/contractual factors particular to each country and its construction industry which would need to be understood by anyone interested in doing business there. Seven chapters are devoted to areas of European Community legislation and policy of special interest to the industry, such as public procurement and standardisation. The final chapter summarises developments and attempts to predict the future for the construction industry in the new, harmonised single market. The book may be read from start to finish, dipped into or used for reference and guidance.

Writing the book has involved aiming at a moving target in difficult and sometimes controversial areas. The pace of European Community legislation and economic and political change in Europe has been hectic. Every effort has been made to ensure the accuracy of information presented but mistakes and misjudgements undoubtedly remain, for which the authors must take the blame. Much of the text was written in early 1991 but was checked for validity in July 1991.

Because of the rapidity of change and because those with a serious interest will need to refer to source texts the authors have provided full reference details and a separate section on sources of information. This provides a great deal of useful information and includes selected addresses and telephone and fax numbers.

The book is written in English for personal reasons and to make it accessible to the widest possible audience. However, many words, expressions and names have been kept in their original language. Translation would generally result in a loss of meaning or charm and in any case why not get used to using the genuine article as soon as possible?

Acknowledgements

The authors wish to thank the many people who have helped in the preparation of this book and the following in particular: Nicholas Forward QC, specialist in European Community law, David White of the Commission of the EC and Richard Grover, specialist in standards and quality management, for their expert advice and suggestions on particular chapters; Dr Haddar, David Robertson, Brian Wood for reading and commenting on draft texts; Gerhard Lohan and Chantal Bruetschy for their friendship and support during and since Nick Spencer Chapman's period of secondment in the Commission and Max Roche of Société Auxiliaire d'Entreprises for encouraging and supporting Christian Grandjean's work on the book. Nick Spencer Chapman would also like to thank his wife and children for their support and for coping cheerfully with the demands made by the book on family life.

List of Acronyms and Abbreviations

ACP	African, Caribbean and Pacific countries (beneficiaries under the Lomé conventions)
AENOR	Asociación Español de Normalisación y Certificación (in Spain)
BAM	Bundesanstalt für Materialforschung und Prüfung
BBA	British Board of Agrément
BERD	see EBRD
BG	Berufsgenossenschaften (in Germany)
BOT	Build, Operate and Transfer (see also 'concession contract')
BRITE	Basic Research in Industrial Technologies for Europe
BS	British Standard
BSI	British Standards Institution
C	see OJ
CAP	Common Agricultural Policy
CARES	Certification Authority for Reinforcing Steels (in UK)
CE	Communauté Européenne
CECA	Communauté européene du charbon et de l'acier (ECSC in English)
CELEX	Law database of the European Communities
CEN	Comité européen de normalisation (European Committee for Standardisation)
CENELEC	Comité européen de normalisation électrotechnique (European Committee for Electrotechnical Standardisation)
CERCLA	Comprehensive Environmental Response, Compensation and Liability Act (in USA)
CIRIA	Construction Industry Research and Information Association (in UK)
CMLR	Common Market Law Reports
COMdoc	A series of documents published by the Commission of the European Communities
COMETT	Community Action Programme in Education and Training for Technology
COREPER	Comitédes Représentants Permanents
COSHH	Control of Substances Hazardous to Health (in UK)
CPD	Construction Products Directive
CSTB	Centre scientifique et technique du bâtiment (in France)

DG	Directorate Générale (of the Commission of the European Communities)
DIN	Deutsches Institut für Normung
EBRD	European Bank for Reconstruction and Development (BERD in French)
EC	European Community (or Communities)
ECHO	European Community Host Organization (for databases)
ECOSOC	Economic and Social Committee (an EC institution)
ECR	European Court Reports (of the Court of Justice of the European Communities)
ECSC	European Coal and Steel Community
ECU	European Currency Unit
EDC	European Documentation Centre
EEA	European Economic Area (EEC plus EFTA)
EEC	European Economic Community
EEIG	European Economic Interest Grouping
EFTA	European Free Trade Association
EIA	Environmental Impact Assessment
EIB	European Investment Bank
EMS	European Monetary System
ENV	European Norms (Standards) for Validation
EOTA	European Organisation for Technical Approvals
ER	Essential Requirements
ERDF	European Regional Development Fund
ESPRIT	European Strategic Programme of Research and Development Technology
EURATOM	European Atomic Energy Community
FIEC	(International European Construction Federation)
GA	Gewerbeaufsicht (in Germany)
GATS	General Agreement on Trade in Services
GATT	General Agreement on Tariffs and Trade
GDP	Gross Domestic Product
GJ	gigajoule (a measure of energy)
GRIM	Groupe Reglémentation Information et Management
HMSO	Her Majesty's Stationery Office (in the UK)
IGCC	Intergovernmental Panel on Climate Change
ISO	International Standards Organisation
L	see OJ
MEP	Member of the European Parliament
MOPU	Ministry of Public Works and Planning (in Spain)
NACCB	National Accreditation Council for Certification bodies (in the UK)
NACE	Nomenclature des activités de la communauté européenne (industrial classification system)
NAMAS	National Measurement Accreditation Service (in the UK)
NATO	North Atlantic Treaty Organisation (OTAN in French)
NEPP	National Environmental Policy Plan (Netherlands)

NL	The Netherlands
OECD	Organisation for Economic Co-operation and Development
OJ	Official Journal L series includes legislation C series includes communications and proposals S (Supplement) series contains tender notices
OPQCB	Office publique pour la construction des bâtiments
PRISMA	Preparation of regional industry for the single market
PSBR	Public Sector Borrowing Requirement
QA	Quality Assurance
RACE	Research and Development in Advanced Communications Technologies for Europe
RNE	Réseau National des Essais (French testing organisation)
S	see OJ
SA	Société Anonyme (*cf* Ltd or PLC in the UK)
SE	Societas Europaea
SME	Small and Medium Enterprise
SPRINT	Strategic Programme for Innovation and Technology Transfer
SRTP	Service de recensement des enterprises de travaux publics
TC	Technical Committee (of CEN or CENELEC)
TED	Tenders Electronic Daily
TIM	Transparent Insulating Material
UEATC	Union européenne pour l'agrément technique dans la construction
UK	United Kingdom
UN	United Nations
UNE	Una Norma Española
VAT	Value Added Tax (TVA in French)
VOB	Verdingungsordnung für Bauleistungen
VOL	Verdingungsordnung für Leistungen
WETT	Water, Energy, Transport and Telecommunications (see Utilities Directive)

Chapter 1

Introduction to the European Community

Membership of the Community affects most aspects of economic life in the Member States. For the construction industry it affects the work to be done, for example through transport and trading patterns, and through harmonisation of environmental standards at levels near to those proposed by the most environmentally concerned Member States. It increasingly affects how work is designed and constructed, and work and safety conditions applicable in the industry. It determines tendering and procurement practices for government contracts, and these have a knock-on effect in the private sector. It already affects aspects of liability and may bring further fundamental and, for practitioners, possibly alarming changes in this area. It even affects the principles of interpretation for regulations and statutes when they derive from Community legislation. Professional qualifications awarded in one Member State must now be recognised in other Member States, subject to certain requirements. The objectives of architectural education around the Community have been harmonised. But most of all, the Community provides opportunities for the construction industry, as well as the threat of increased competition. The economic environment is going through a period of rapid change. Companies which learn, plan and anticipate will prosper, while those which pay little attention to developments may disappear.

This chapter sets out the fundamentals of the Community – its members, formation and objectives. It describes the programme for the completion of the internal market, popularly referred to as '1992', and the intensive economic study, the 'cost of non-Europe', which attempted to identify the savings and growth which could be expected from the completion of the internal market. This book is not written exclusively for a United Kingdom readership, but reference to the British experience within Europe and to practice in the UK is made from time to time. The section on the 'UK in the Community' (see p 13) outlines the UK's ambivalent relationship with the Community and shows how the European market, including the construction market, has become more important for the UK. At the time of writing, traditional overseas markets for the UK construction industry have dwindled and the construction market at home is in recession. This, together with the publicity for '1992', has led to increased interest and activity in Europe. Even so, British business remains relatively poorly informed about the Community and other Member States. It also suffers the handicap of a national weakness in languages.

Finally, this chapter considers some of the more important aspects of the Community's external relations, GATT, EFTA and Lomé. GATT, the General Agreement on Tariffs and Trade, is important for trade and economic growth. Current negotiations aimed at extending the main agreement, and a subsidiary agreement concerning public procurement, to services could lead to new opportunities for construction and consulting abroad. Policies in many areas are co-ordinated between the Community and EFTA, the European Free Trade Association, of which the UK used to be a member. Negotiations between the Community and EFTA are, in practical terms, likely to extend the internal market to the six EFTA members in an 18 country 'European Economic Area'. The EFTA member countries already present opportunities and threats for Community industry. Under the Lomé conventions the Community provides grants for aid and assistance for many developing countries, primarily countries with former colonial links with Community Member States. The contracts awarded in connection with the conventions may interest consultants and contractors seeking work in developing countries.

Some readers may prefer to skip some of this and the following chapter, which deals with the Community institutions, and move straight on to chapter 3 on the construction industry in the European Community. Following chapters presume a knowledge of certain topics such as the organisation and authority of the Commission and the Council of the European Communities. However, the index may be used as needed to guide the reader to particular topics.

Members, origins and objectives

Member States

The European Community consists of 12 Member States:

Belgium	Italy
Denmark	Luxembourg
France	The Netherlands
Germany	Portugal
Greece	Spain
Ireland	United Kingdom

Table 1.1 Data on EC Member States, certain regions and selected other countries

	GDP 1988 ECU billion**	Growth 1983–88 %	Gross fixed capital formation % of GDP	Population 1988 1000s	Area 1000 sq km	Population density persons /sq km	Estimated population 2000 1000s	Local currency to ECU 1989
EC 12*	4032.2	2.8	20.0	324611	2253.3	144	330150	
Belgium	127.0	2.2	17.8	9902	30.5	324	10163	43.3806
Denmark	91.0	2.1	18.0	5130	43.1	119	5080	8.04928
W Germany	1017.6	2.5	19.9	61450	248.7	246	58217	2.07015
Greece	44.4	2.1	17.4	10004	132.0	76	10600	178.84
Spain	287.9	3.6	22.5	38765	504.8	77	41193	130.406
France	804.2	2.1	20.1	55884	544.0	102	57764	7.02387
Ireland	27.5	3.0	17.0	3538	68.9	51	3413	0.776818
Italy	701.8	3.0	19.9	57452	301.3	190	56400	1510.47
Luxembourg	5.6	4.1	24.2	375	2.6	143	389	43.3806
Netherlands	193.2	2.3	21.4	14760	41.2	356	15484	2.33526
Portugal	35.3	2.8	26.8	10287	92.1	112	11150	173.413
UK	696.6	3.6	19.2	57065	244.1	223	59918	0.673302
UK S East				17317	27.2	636		
UK Scotland				5112	78.8	65		
E Germany				16706	108.2	154		
Sweden	153.8		19.7	8436	449.9	19	8322	7.09938
Austria	107.7		23.5	7595	83.9	90	7461	14.5695
Soviet Union				283682	22402.2	13	307737	
USA	4076.0	4.4	17.1	246329	9372.7	26	266194	1.10175
Japan	2403.6	4.4	30.5	122613	372.3	329	129105	151.938

* excludes East Germany
** i.e. 1000 million (= milliard in French)
Source: Eurostat

According to Michael Toner's 'Bluffer's guide to the EEC' the 12 have only three things in common:

> 'friendships (which are rare), rivalries (which are universal) and hatreds (which are common) . . . the one exception being Luxembourg, since not only is it not big enough to rank as anyone's enemy, rival or even friend – but no one knows where it is anyway. . . . They differ in language, religion, outlook and economic performance. Most of them have spent the last several hundred years being excessively rude about their neighbours, when they weren't actually attacking them.'

Before the unification of Germany in October 1990 the population of the Community stood at 324 million, and its Gross Domestic Product made it the largest market in the world. East Germany has added 16.7 million to the EC population, although much less in relative terms to GDP.

Expansion

The Community is likely to grow, possibly to as many as 20 Member States by the end of the century. Malta, Turkey, Cyprus and Austria have all submitted applications. Malta's membership is likely to pose few problems. It has had an association with the EC since 1971 and its stage of development is similar to some Member States. Austria, already on a busy transit route between Germany and Italy, would fit in easily, with income per capita higher even than that of Germany. The processing of the Cyprus application could be complicated by its division into Greek and Turkish regions following a Turkish invasion in 1974.

Turkey, a country of some 55.5 million inhabitants, has campaigned strongly for membership. However, its poor human rights record and very low standard of living, which would require substantial injections from the various EC funds, may make it difficult to accommodate. Its application has been rejected once already, in December 1989. Sweden, which already participates actively in EC industry, would be a natural member and has come under pressure from its industrial sector to join, partly to obtain more influence in the decision making processes. The government appears to be enthusiastic, but has as yet concentrated on efforts to bring the European Free Trade Association (EFTA) and the Community closer in the creation of a 'European Economic Area'. Norway, another member of EFTA, rejected membership in a referendum in 1973 but is considering applying again. The other members of EFTA – Iceland, Finland, and Switzerland – may follow.

The democracies emerging from the eastern bloc, particularly Poland and Hungary are also interested in eventual membership, though their economies will need substantial changes before entry could be considered. In any case, the Community has decided that the completion of the internal market should take precedence over the admission of new members. None will be admitted until 1993 at the earliest.

Origins of the Community

A number of institutions sprang out of the post Second World War determination that Europe should never again engage in mutual destruction. Three main institutions (treating the European Communities as one) were established. The first was the Paris based Organisation for European Economic Co-operation which the USA helped create to co-ordinate the Marshall Plan in which US funds were used with great success to reconstruct Europe. The US and Canada later joined the organisation and it changed its name to the Organisation for Economic Co-operation and Development (OECD).

The second institution was the Council of Europe (easy to confuse with the Council of Ministers of the European Communities). The Council of Europe is based in Strasbourg, in France. Its aim was intergovernmental co-operation in various fields and is best known for the 1950 European Convention for the Protection of Human Rights and Fundamental Freedoms and the Court of Human Rights (more scope for confusion, this time with the Court of Justice of the European Communities which is based in Luxembourg). Complicating matters further, the Council of Europe owns the building in Strasbourg where the European Parliament, a Community institution, holds its plenary sessions.

Finally, there was the creation of the European Communities, comprising the European Coal and Steel Community (ECSC), the European Atomic Energy Community (EURATOM) and the European Economic Community (EEC). These differed from the first two institutions, because from the beginning their explicit aim was integration.

The European Community

Winston Churchill had first suggested a United States of Europe in

1946 and in 1950 the French Foreign Minister Robert Schuman (hence Rondpoint Schuman, the roundabout adjacent to the main EC Commission buildings in Brussels) proposed that the coal and steel industries of Germany and France should be subject to co-ordination under a single authority, in part to avoid the friction which had in the past lead to war.

The European Coal and Steel Community (treaty signed in 1951 on behalf of the original six – Belgium, France, Italy, Federal Republic of Germany, Netherlands and Luxembourg) was successful enough to lead to the creation of the other two communities (EURATOM and EEC) under treaties signed in March 1957. The Community used to be referred to as the EEC or the Common Market, but in 1987 it officially renamed itself (some might prefer to say 'themselves') the European Community; as described later, the bodies established under the treaties had been merged somewhat before, in 1965. Even now though, the bodies officially retain 'of the European Communities' in their names, for example, the Commission of the European Communities. This is Europe imitating the three in one and one in three, but is logically difficult to handle.

The nature of the European Community

The European Community is a body founded on law. The Member States undertake to allow certain decisions to be taken by the institutions of the Communities and to accept the rulings of the Court of Justice. In doing so they accept some loss of sovereignty. With the signing of the Single European Act in 1986 further sovereignty was ceded to the Community, in particular when qualified majority voting (weighting of votes giving more power to the larger countries) was accepted for a wide range of matters. The issue of sovereignty is sensitive in some quarters, particularly in the UK, although there remains the ultimate remedy of withdrawing from the Community. In that sense there has been no true loss of sovereignty.

A prime objective of the Community is union. The first preamble to the Treaty of Rome (the EEC treaty) reads: 'Determined to lay the foundations of an ever closer union among the peoples of Europe . . .' while article 1 of the Single European Act reads: 'The European Communities and European Political Co-operation shall have as their objective to contribute together to making concrete progress towards European Unity.' How rapidly monetary and then political union progresses remains to be seen.

The Community already has some of the characteristics of a nation state, and may be unique in being founded on law rather than on force.

1992 – the single market (also known as the internal market)

The White Paper on the completion of the internal market

The Treaty of Rome expressed intentions which would have meant a high degree of integration by the end of the 12-year transitional period ending on 1 January 1970. Progress was slow and during the 1970s came to a virtual standstill, with much cynicism becoming evident. The gradual loss of competitiveness in comparison with Japan and other countries, among other factors, led to renewed calls for economic integration which gathered momentum during the early 1980s.

In December 1982 the European Council (not the Council of Minsters but the twice-yearly meeting of the Heads of State and the President of the Commission) instructed the Council of Ministers to 'decide, before the end of March 1983, on the priority measures proposed by the Commission to reinforce the internal market'. In June 1984 it asked for 'measures which could be taken to bring about in the near future . . . the abolition of all police and customs formalities for people crossing intra-Community frontiers' and a few months later it agreed that steps should be taken 'to complete the internal market, including the implementation of European standards'. The objective of completing the internal market was to create a Community without barrier to the free movement of goods, services, people and capital, in which firms and individuals could establish in whatever Member State suited them. Finally, in March 1985 it called on the Commission 'to draw up a detailed programme with a specific timetable before its next meeting'. The 'White Paper on the completion of the internal market' [1.1] was the result. Some 300 measures, since reduced to 279, were described, with a programme for each one.

The Single European Act and '1992'

To accelerate moves towards European integration, and to widen its scope, the European Council decided in June 1985 to create the legal instrument now known as the Single European Act [1.2]. This act was signed in February 1986 and came into force on 1 July 1987. It takes the form of a treaty which introduces changes, including substantial new additions, into the Treaty of Rome. One of the new articles introduced by the SEA is Article 8a:

> 'The Community shall adopt measures with the aim of progressively establishing the internal market over a period expiring on 31 December 1992, in accordance with the provisions of this Article . . . The internal market shall comprise an area without internal frontiers in which the free movement of goods, persons, services and capital is ensured in accordance with the provisions of this Treaty.'

One of the important changes brought about by the Single European Act was the introduction of a form of majority voting, called qualified majority voting, for a variety of measures concerned with the creation of the internal market. Prior to that most measures had been held up or blocked in the Council because the voting procedure required unanimity.

The single market programme

The programme has three main objectives:

(1) The creation of a single market through the removal of physical, technical and fiscal barriers. With over 320 million inhabitants it would have a purchasing capacity larger than any other single market in the world.
(2) To ensure the technical vitality of the market through programmes promoting advanced technology such as ESPRIT (the acronym of a programme concerned with research and development of information technology), RACE (telecommunications), BRITE (Basic Research in Industrial Technologies). Related to this was the objective of barriers to trade caused by differing, or weak 'intellectual property rights', i.e. patents, trademarks and copyright.
(3) To enhance economic flexibility by eliminating barriers to the movement of people, goods, services and capital. Enhanced scale of production and specialisation would enhance competitiveness. Associated with this would be the control of anti-competitive behaviour by enterprises. Also, Member States would be prevented from unfairly protecting national companies and supporting unfair competition by subsidies or other forms of support.

Though not part of the White Paper, the concern to limit economic damage to particular regions or disadvantaged groups caused by new competition, has lead to strengthening of social and regional policy. Social policy is concerned largely with conditions of employment and access to training. Regional policy has tended to focus on providing funds to promote the physical infrastructure needed for development. It has also meant providing funds, training and information for small and medium enterprises (SMEs).

Aspects of the internal market programme

Removal of physical barriers

Measures were proposed to remove controls, particularly at frontiers, which complicate travelling and transporting goods between Member States.

Removal of technical barriers and free movement of goods

This section was concerned mainly with developing European Standards, including those for construction and construction products, and mutual recognition of tests and certificates. The developments in standardisation were to build on the then recent 'new approach'. This referred to a decision to abandon the previous, and largely unsuccessful, approach of trying to agree detailed technical requirements. Instead the intention was to introduce in each product area directives specifying minimum essential requirements, mainly concerned with health and safety. The European standards organisations (CEN and CENELEC – see chapter 9) were then to be given the task of developing or adopting specific standards. There was also reference to Eurocodes in relation to structural design. Directives were called for on the safety of buildings (resistance to earthquakes, other accidental damage, prevention of progressive collapse), reliability and durability of buildings, test methods for reaction to fire and resistance to fire of building materials and components, the prohibition of the introduction of new or modified building regulations at local government levels for five years, and on revolving cranes.

In the longer term (1987–92) directives were required on safety and reliability of buildings, fire safety in public buildings and hotels, insulation products, and the safety and reliability of special structures (masts, towers, long bridges, drilling platforms and water-retaining structures). The actual development in this area has followed a different pattern. One broadly based directive, the Construction Products Directive [1.3] had been adopted. This specifies the 'essential requirements' (ERs) which must be capable of being met by structures incorporating construction products. For example, ER1 covers mechanical resistance and stability and ER2 covers safety in case of fire. The Construction Products Directive and related issues are discussed in chapter 9. The White Paper also called for the development and introduction of a model code for building regulations (based on UN-EC models).

Public procurement was given much prominence with improvements required in the Public Works Directive, which applies to tendering procedures and specifications for building and civil engineering contracts in the public sector. Directives were required to extend the application of rules and procedures to the purchasing by entities in the water supply, energy, transport and telecommunications sectors and to the procurement of services, including architectural and engineering consultancy services.

Free movement of labour and the professions

This section called for proposals on mutual recognition of training and qualifications as well as measures intended to make movement for migrant workers and others, such as students and retired people, easier.

Common market for services

This section was concerned with banking, insurance and securities. Though of little direct concern to the construction industry some of the administrative complications of arranging insurance and transferring money between Member States could be eased by these proposals. This section also dealt with opening markets in transport, particularly road and air transport, new technologies and services such as broadcasting and cable transmission.

Free movement of capital

This section was aimed at removing restrictions on cross-border financial transactions, particularly those involving mortgages and securities.

Creation of suitable conditions for industrial co-operation

This section dealt with company law, including the regulation for a European Economic Interest Grouping (EEIG), a pan-Community entity making cross-border ventures between companies more straightforward, and other matters such as takeover bids, intellectual and industrial property and taxation.

Application of Community law

This listed a number of 'communications' (discussion documents addressed to the Council of Ministers) to be made by the Commission in relation to obstacles to movement of goods, and dealt with competition policy and state aids.

Removal of fiscal barriers

This section was concerned with VAT and excise duties. The intention was to bring about harmonisation of rates and to reduce or remove the complications when cross-border movements of goods or services take place. At present the system for remitting the taxation to the correct national authorities depends on time-consuming paperwork, particularly at borders.

Progress on the internal market programme

In December 1989 Jacques Delors, the Commission President, reported that the Commission had drafted 274 out of the 279 measures in the internal market programme, and that by the end of that year 60 per cent of the measures would have been adopted. By January 1991, 200 out of the now 282 proposals (71 per cent) had been adopted. However, progress in some of the key areas has been slow, particularly on the harmonisation of tax rates, the removal of border controls, and liberalisation in financial services. Not strictly part of the original programme, but seen by Mr Delors as a vital accompaniment is the social dimension, and on this there is concern in some quarters about the lack of progress. The areas on which attention may be focussed are wages and working conditions (including determination of a fair wage, procedures for mass redundancies, and conditions in sub-contracting), industrial relations (particularly worker participation), health and safety, and access to training.

Despite these particular problems, the excellent progress on the completion of the internal market has surprised many. The internal market is already partly here, but in some of the more difficult areas progress is still to be made, and often the end result will not be quite as envisaged in the original vision.

Perhaps of greater significance is the change in business attitudes. The '1992' campaign has been successful in raising awareness and interest in the Community. There has been a significant rationalisation and restructuring of certain industries, generally through mergers, acquisitions and various forms of co-operation across borders. As much, if not more, notice has been taken of the internal market programme by countries outside the Community. They also stand to benefit from being able to deal with the EC as a single market instead of with 12 separate countries. Secondly, they have been concerned that the breaking down of the internal barriers would be accompanied by an increase in protectionism, often referred to as 'fortress Europe'. The response has been a considerable increase in 'inward direct investment' with the Japanese being amongst the leaders. How many of these developments would have taken place without the internal market programme, as a result of the increasing globalisation of markets and increased competition is debatable.

Other developments, stemming in part from the increased attention on the EC which the internal market has brought, are the calls for increased political co-ordination, and even the creation of an EC military force. Whether the EC can maintain the momentum towards integration in the face of recession and mounting external pressures, such as the developments in eastern Europe, the disintegration of the Soviet Union and war and instability in the Middle East is hard to predict.

The cost of non-Europe

In early 1986 (after the issue of the White Paper on the completion of the internal market) the Commission launched a major study, undertaken by W.S. Atkins Management Consultants in association with Eurequip SA-Roland Berger and Partner-Eurequip Italia on the 'cost of non-Europe', popularly known as the Cecchini Report after the chairman of Commission's committee guiding the research. The 13-volume study was summarised and popularised in a 110-page paperback, *The European Challenge: 1992 – the benefits of a single market*. The study concluded that even without the effect of medium-term economic measures to accompany the completion of the internal market, the internal market would boost the Community's GDP by 4.5 per cent. In employment terms the overall effect was estimated to be 1.8 million additional jobs. If the medium-term economic measures were implemented, 5 to 7 per cent additional non-inflationary growth was forecast, with two to five million new jobs being created. Table 1.2 shows the estimated benefits (without the accompanying economic measures).

Table 1.2 Estimated benefits to be derived from the internal market

Cause	Relative change in GDP	Effect on consumer prices
Eliminating customs formalities	0.4%	-1.0%
Opening up public procurement	0.5%	-1.5%
Liberalising financial services	1.5%	-1.4%
Restructuring of industry leading to economies of scale and effect of increased competition on prices (also includes benefits from standardisation)	2.1%	-2.3%
Total mean value	4.5% ±1.2%	-6.1% ±1.6%

The UK in the Community

Churchill had suggested a United States of Europe, but the UK refrained from becoming a member of the European Coal and Steel Community in 1951. By 1959 the UK had been instrumental in establishing the European Free Trade Association (EFTA), which then also included Norway, Sweden, Denmark, Austria, Portugal, Iceland, and Switzerland, with Finland as an associate member. Membership of EFTA, which had no institutions equivalent to the Court of Justice or the Commission with powers over member states, involved no loss of sovereignty. But by 1961 the UK had seen the success of the European Communities (plural at that time) and made its first application to join. Both this application and the subsequent one in 1967 were blocked by President de Gaulle of France, and it was not until he resigned that the negotiations for membership were able to continue under the strongly pro-Community UK Prime Minister, Edward Heath. The Treaty of Accession was signed on 22 January 1972, and after a referendum showing a clear majority for entry, the UK joined on 1 January 1973.

Even now, elements in British society and politics, both on the right and the left, are equivocal or even hostile to the Community. Under the leadership of Mrs Thatcher from 1979 until 1990 the UK gained a reputation as being an extremely reluctant European, blocking many of the initiatives apparently supported by other Member States. Her Bruges speech at the College of Europe on 20 September 1988, in which she said that it would be folly to mould the national identities of the member states into 'some sort of identikit European personality' and that 'a United States of Europe was out of the question because people owed allegiance to their own nation states, not some wider entity' cemented friendships in the anti-EC lobby while it outraged many others. A senior EC official was quoted as saying that her comments were 'outrageously and unrelentingly negative', while a Conservative member of the European Parliament on the other hand welcomed her speech and said that 'as long as Mrs Thatcher lived and breathed in Downing Street there would not be a United States of Europe'.

At a practical level the UK was less obstructive. Mrs Thatcher signed the Single European Act in 1987, which allowed qualified majority voting (see p 21) on a wide range of issues. In Brussels the UK representatives provided steady and constructive support in committees, much more constructive than that of some of the avowedly more pro-integrationist states. The UK has one of the best records in implementing proposals, and the UK is more open to foreign competition than almost any other Member State. Mrs Thatcher particularly supported the opening up of the internal market and the removal of barriers which coincided with her own free-market philosophy. It was the notion that this should be accompanied by the social dimension and that the ultimate objective was union, first of all monetary and then political, which concerned her.

Commercially it has proved difficult for the UK to shift its focus from the former colonies and other areas of influence, where English was the national language or widely used, and where the English system of law applied. But trading patterns have changed dramatically. Trade with other Community members increased between 1958 and 1985 from 22 per cent of the total to nearly 50 per cent. The construction industry in the UK had engaged in much overseas work, but primarily in the Commonwealth countries and other areas of former influence such as the Middle East and Africa. By 1990, however, 34 of the 100 largest UK construction companies were reporting work undertaken in the EC, an increase from 22 (which then included all western Europe) in the previous year. Other figures of interest were that in 1990 only 22 were still working in the Middle

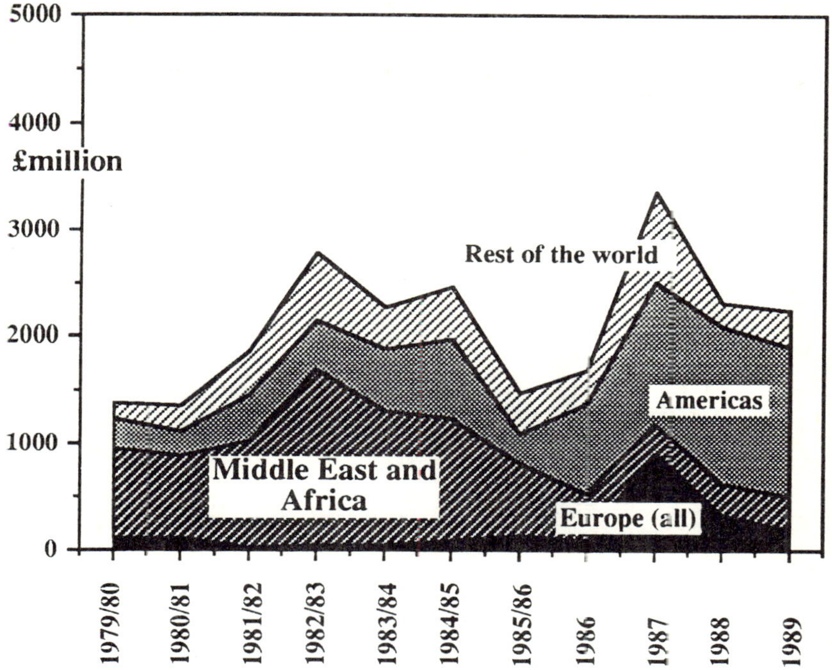

Figure 1.1 Overseas orders won by UK construction companies and their subsidiaries at current prices.

East, while 11, five more than in 1988, were working in eastern Europe.

Is then Europe a major export market for British firms and their subsidiaries? Figure 1.1 shows how very small it is. It also shows the decline in value of overseas orders and that the dramatic decline in Middle East and African work has been partially replaced by orders obtained in the Americas; the latter being accounted for mainly by the takeover of US firms. Apart from a blip in 1987 the value of foreign orders in Europe remains minimal.

Figure 1.2 shows how foreign work fell during the period 1982/83 to 1989 (the 1987 figure is distorted by the Channel Tunnel contract) while UK domestic orders grew particularly strongly until 1988. In 1990, the UK construction industry, which for the two or three years previously had been rushed off its feet, suddenly had to cope with an alarming collapse in orders and a sharp recession (not the 'soft landing' spoken of by Nigel Lawson in 1989/90 when still Chancellor of the Exchequer).

This drop in UK workload, combined with the interest whipped up

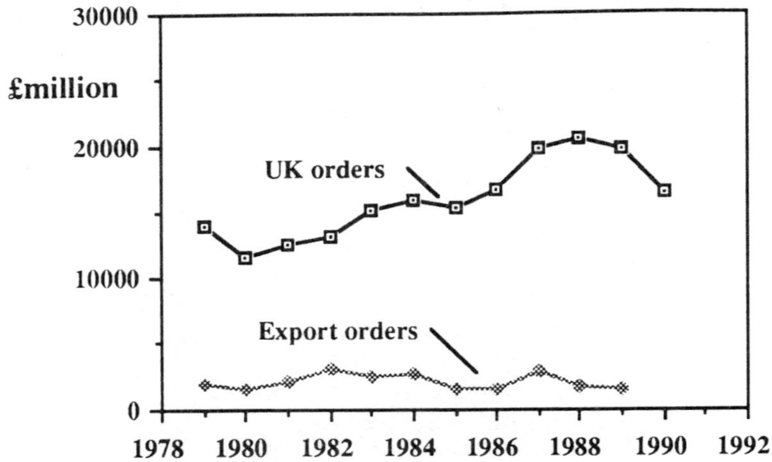

Figure 1.2 Orders obtained by UK construction companies and their subsidiaries in the UK and overseas at 1985 prices.

(*Source:* HMSO Housing and Construction statistics.)

in European markets by the Department of Trade and Industry, the National Economic Development Office, the professional institutions and others, accounts for the sharp increase in interest in these markets. The activity of the top 100 companies shows this, as does the coverage in the press and the growth in the number of conferences about European markets. Another factor is, of course, the growth in legislation and threatened legislation originating from Brussels, as well as the development of standards at European level.

Much of the reporting has been alarmist. For example, in 'Construction: restructure to win', a memorandum dated 9 January 1991, by Sir Christopher Foster, Chairman of the National Economic Development Council, Construction Industry Sector Group, the industry was warned: 'If UK firms, including designers, suppliers and contractors, do not respond to these challenges, they will lose market share and end up with the lower value added elements of the construction process. There is a danger that British workers will become the "navvies of Europe".' The document also reported a number of chilling statistics, such as very low and falling investment to output ratio, high building costs in comparison to the USA, Germany, France and Italy despite the lowest remuneration per employee amongst that group. The UK was also shown as having the lowest output per head in the same group (64 per cent of the US figure).

There is also concern about standards being developed which are geared towards the practices and capabilities of other Member States. 'UK losing out on Euro standards' proclaimed a title above an article in *New Builder* of 16 August 1990 which began 'Powerful foreign interest groups are manipulating the drafting of European standards'. The article was about a report produced by the Building Services Research Information Association's Eurocentre in Bracknell, Berkshire, which states that France and particularly Germany are beginning to dominate the drafting committees for standards.

A clear indication of UK interest in the Community is the increase in cross-border takeover activity by British companies, which during 1990 increased by 57 per cent to ECU 5.9 billion, and which accounted for 35 per cent of total international spending on acquisitions. The USA however remained the most popular destination with ECU 8.9 billion, even though this represented a considerable reduction from 1988 (ECU 27.4 billion) and 1989 (ECU 14.5 billion). Food, drink and tobacco was the most active sector, but this was followed by construction and building products. France was the most popular destination with 62 deals, followed by Germany where 51 had been agreed [1.4].

Some interest was also shown by the UK in the eastern European countries. A trade mission was led by the Construction Minister, Mr Michael Spicer, who commented 'there is a tremendous amount of work to be undertaken and it is unlikely that we shall see another chance such as this before the end of the century. Although they will represent a challenge, the opportunities for the UK construction industry are undoubtedly real.'

Despite the interest, the amount of work carried out by British construction companies and their subsidiaries remains very small. The British, traditionally weak in languages, have proved reluctant to make the necessary commitment to markets on the continent and still remain suspicious of, or even hostile to, many initiatives emanating from Brussels.

There appears to be some alarming ignorance in the UK about the Community. A study of business executives presented in mid-1990 [1.5] reported 'ignorance, insularity and parochialism', with only 37 per cent knowing that the headquarters of the Commission was in Brussels, that Bonn was the capital of Germany and only a third being able to say 'Good morning, my name is . . .' in German. Despite the ignorance, 44 per cent believed that they were equipped for the challenge of the single European market. Two thirds of executives described the single market as the most important business event of the decade, but 53 per cent thought that it would have little or no impact on their businesses.

GATT, EFTA and Lomé

The Community is party to a wide variety of international agreements and has ambassadors in many countries. Brussels accommodates not only embassies to Belgium, but to the European Community as well. The most significant of the agreements, which have some relationship with the construction industry, or could have in the future are described below.

General Agreement on Tariffs and Trade (GATT)

The General Agreement on Tariffs and Trade is an international organisation set up in 1948 with a secretariat in Geneva. Signatories to the agreement now number over 100 and include many eastern European nations such as Poland, Czechoslovakia and Hungary. However, the main players in the negotiations are the USA, Japan and the EC, in the form of the Commission, which negotiates on behalf of its Member States. The Commission delegate speaks during the negotiations, but the negotiating position is agreed in freqent Council meetings, held in the Council's Geneva chambers. Each Member State also has delegates who may attend the negotiation sessions. The agreement deals with barriers to trade, reduction in import tariffs and quotas and the abolition of preferential trade agreements. For the moment the agreement is restricted to goods, but may be extended to services, including construction, in what is referred to as GATS (General Agreement on Trade in Services). Another very significant aspect is the possible extension of the Agreement on Government Procurement, to which a small number of GATT members are signatories, including the USA, Japan and the EC. The extension would cover government procurement of public works, and is discussed further in chapter 6.

The agreement has been extended through a series of negotiation rounds, the latest of which is referred to as the Uruguay round, after the opening meeting of ministers held in Punta del Este, Uruguay, in 1986. The round was due to reach a new agreement at the end of 1990, but concluded in acrimonious stalemate, with many nations including the USA blaming failure on the EC's determination to keep its protectionist Common Agricultural Policy in place. In fact, the USA also carries a burden of responsibility. It was determined to maintain its own freedom to retaliate unilaterally against countries which it alleges are not sticking to the agreement, and in 1990 changed policy on opening up markets in financial services, transport and telecommunications. The developing countries are keen to see the construction markets of the richer countries, particularly the Community, USA and Japan, opened up, with special measures to allow them to carry out work in other countries using their own temporarily imported labour. Commentators expected the developed countries to grant some concessions on construction in return for opening up by the less developed countries in the higher technology areas such as telecommunications. At the time of writing the negotiations have resumed in Geneva, but there are still major stumbling blocks.

The GATT agreement has been a very significant contributor to the growth in world trade since the Second World War and a breakdown in the system could lead to fragmentation of the world market into separate trade blocks with serious consequences for the prosperity of the Community. Agricultural production in the Community is a minor component of Gross National Product but defence of the Common Agricultural Policy could threaten the GATT and result in widespread disruption to world trade.

European Free Trade Association (EFTA)

It will be recalled that the UK was instrumental in establishing the European Free Trade Association (EFTA) before leaving to join the Community. The members of EFTA are now Norway, Sweden, Austria, Switzerland, Finland and Iceland – a group collectively dwarfed by the Community. EFTA is a much looser body than the Community and has no equivalent to the Court of Justice which has supremacy over national law. Also the agreement only extends to trade amongst its members, without having a common policy on trade across its 'external borders' (an interesting concept given the spread of its members). Well advanced negotiations between the Community and EFTA are intended to lead to the creation of a European Economic Area (EEA), in which the EC single market freedoms relating to capital, goods and services would be extended to EFTA. This would in some respects be a 'single market' of 18 countries. The EEA agreement would probably include measures on public procurement. Already the EFTA countries participate in standards setting through CEN and CENELEC (see chapter 9).

Lomé

The Lomé conventions date back to 1975 when Lomé I was agreed between the EEC (as was) and, then 46 developing countries in Africa, the Caribbean and the Pacific (known as the ACP states), in Lomé, the capital of Togo. Under the convention all ACP industrial exports and most agricultural exports to the Community are free of duty. The interesting part for the construction industry is that the Community also agreed to provide grants for financial and technical aid, much of which is related to civil engineering consultancy and construction.

Lomé III was agreed in 1984 and covered 66 ACP countries, with eight non-EC Mediterranean countries signing co-operation agreements. Under it a package worth 8500 million ECU (approx. £6000 million) was made available for the 1985–90 period. Lomé IV has recently been agreed, with a considerable increase in funding. Many civil engineering consultancy and construction contracts are available under Lomé funding. About 150 invitations to tender are issued each year with about 50 per cent of the contracts being works contracts, 30 per cent supply contracts and the remainder service contracts including training programmes, feasibility and technical studies and contract supervision.

Lomé funds are administered by Directorate Générale VIII of the Commission (see chapter 2 for information about the Commission) which every two months publishes a list of the contracts available in *ACP – EEC Courier*. Notices are also published in the S series of the EC's *Official Journal* and in TED, *Tenders Electronic Daily* [1.6]. The procedure is that DGVIII selects a number of suitable firms whose names are sent to the country in which the project is to be carried out. The authorising officer in the country is responsible for the selection of the firm, and may add other names to the tender list. One problem participants face is that there is an unofficial quota system limiting the proportion of contracts won by national firms to the proportion of funds contributed by that country. One solution is to form a relationship with a firm in another Member State whose quota has not yet been filled. The northern Member States apparently reach the quota limit rather sooner than those in the south. Another factor influencing selection may be bribery in the recipient country.

For UK firms, the world aid section of the Department of Trade and Industry can provide information. The national permanent representative's office in Brussels may also be helpful.

Chapter 2

Institutions of the Community

To understand the Community and know how to influence the decision making process or obtain information or funding, a knowledge of Community institutions is indispensable. The main institutions of the Community are the Council of Ministers, the Commission, the European Parliament and the Court of Justice. Originally each of the three communities (coal and steel, economic and atomic energy - see Chapter 1) had its own institutions. Conventions merged the institutions to produce the Court of Justice, the Economic and Social Committee (1957), the Council of the European Communities and the Commission of the European Communities (1965).

The Council of Ministers

The Council is the Community's decision-making body. It is the body which takes a 'common position' on proposals for directives issued by the Commission and at the end of the day 'adopts' directives, which are then 'communicated' to Member States who are obliged to transpose them into national law within a given period. It is the Council, not the Commission, which has the main Community legislative powers.

Most measures concerned with the single market, together with those on the environment and health and safety, amongst others, are subject to 'qualified majority voting' in the Council. A qualified majority requires 54 votes in favour out of a total possible of 76 which are allocated as follows: ten each for Germany, France, Italy and the UK, eight for Spain, five each for Belgium, the Netherlands, Greece and Portugal, three for Denmark and Ireland and two for Luxembourg. This gives Germany the least say per head of population, while Luxembourgers get apparently disproportionate power! No wonder some Scots would like separate Member State status. Under the qualified majority voting procedure a measure can be blocked by two of the larger states plus one other so long as it is not Luxembourg.

Decisions, such as adoption of directives and regulations, are taken by meetings of the ministers of the appropriate departments in the Member States. But the term 'Council' also embraces the working groups of national officials and the 2000 staff of the institution. Incidentally of the 2000, 1800 are concerned with translation and interpreting. Translation is of documents, while the highly paid interpreters are mainly concerned with simultaneous translation at the many meetings proceeding each day: an awesomely difficult task undertaken from a series of darkened, glass fronted chambers around the meeting rooms.

The chairing (in French a chairman is *le Président*) of the meetings of ministers and all the working party meetings is by the representative of whichever Member State holds the Presidency of the Council. This rotates on a six-month cycle, currently (January to June 1991) with Luxembourg. The Member States tend to compete with each other on achievements during their period of presidency.

The bulk of the discussion of and negotiation about proposals is handled by working parties, with representative individuals or teams from each Member State. These either come from the Permanent Representative's office which each Member State maintains in Brussels, or are sent for the meetings from national ministries. The ministries normally seek opinions widely from within government, industry and elsewhere about proposals during the course of discussions.

After meetings of the working parties, which may take place over many months, discussion moves up to the COREPER meetings in the Council. These are the COmité des REprésentants PERmanents referred to above, and once they have reached agreement or impasse it is passed onto the ministers themselves.

The main Council building in Brussels, on the impossible to cross five-lane one-way rue de la Loi, is adjacent to, and down-hill from, the Berlaymont building, the headquarters of the Commission. It is strangely anonymous with no identification other than the number. Many of the meetings of the ministers themselves take place at the Council's offices in Luxembourg.

The Commission of the European Communities

The Commission is the executive branch of the Community. It has obligations imposed on it under the treaties to *propose* policy and legislation, most of which must then be passed onto the Council for eventual adoption. The Commission also carries out Council instructions and supervises the implementation of Community policy. It ensures that the Member States and others comply with Community law and in certain cases has the power to impose fines, some of which may be very large and against which there is only two months in which to appeal. In some cases it can impose interim measures (in the legal sense) which must be complied with pending a hearing in the Court of Justice. Finally it negotiates international trade agreements, such as the GATT agreement or the Convention on CFCs (chlorofluorocarbons) on behalf of the Member States.

The Commission itself is composed of 17 members, two each from Germany, France, Italy, Spain and the UK and one each from the remaining Member States. Commission officials are careful to distinguish between matters which the Commissioners themselves have agreed and decisions, documents and opinions of the Commission services. Each Commissioner has responsibility for one or more areas of Community policy. The Commissioners, though appointed by Member State governments (for a four-year period), undertake to serve the interests of the Community and not their Member State, and most are successful in doing so. The present President of the Commission is Jacques Delors of France, the first president to be appointed for a second term of office. Though said to be overbearing and difficult to work with, he has had the vision and energy to reinvigorate the Community and is widely regarded as the inspiring force behind the '1992' programme and the Single European Act. He is also strongly committed to rapid European Monetary Union, with a single currency, and to political union. One of the UK Commissioners during the four-year term of office which expired at the end of 1988, was Lord Cockfield, who is given great credit for determining and masterminding the details of the 1992 programme. Mrs Thatcher as Prime Minister was known to dislike Mr Delors and was suspicious of Lord Cockfield, considering that he had 'gone native' in Brussels, adopting a European vision.

It is customary for each Commissioner to have a personally chosen team known as a *cabinet* (pronounced in the French way) which is headed by a *chef de cabinet*. Those new to Brussels may be surprised to be introduced to a Commissioner's chef but the position is influential!

The Commissioners occupy the 13th floor of the building which is used as the symbol of the Community, the Berlaymont building, the one with elegantly curved facades and a row of Member State flags on stainless steel flagpoles outside, 200 rue de la Loi. Shaped a little like the fuselage and swept back wings of an aeroplane the building, which was built in the 1960s, actually belongs to the Belgian government. The Belgians have tried to sell it to the Community for around ECU 160 million. However the building is rather dowdy inside, poorly ventilated, and was constructed with a considerable amount of asbestos insulation. There is constant monitoring of asbestos levels in the building and occasional evacuations when the safety threshold is exceeded. An ECU 95 million operation to remove the asbestos is planned for 1994, but has been questioned by some on the basis of a report by the USA's Environmental Protection Agency which concludes that removing asbestos is often more dangerous than leaving it in place. The building also suffers from a Belgian idiosyncrasy which would probably not be permitted under UK fire regulations, narrow stairs sometimes hidden from view; Belgian architects apparently expect everyone to use lifts.

Only about 3000 Commission staff are housed in the Berlaymont. The remainder are scattered around the western part of Brussels in over 30 buildings. Another part of the Commission, mainly concerned with statistics and publications is located in Luxembourg about three hours drive away. There are about 10 000 staff working for the Commission which makes it comparable with a medium sized ministry in a Member State. It also has to cope with the burden of nine official languages, though formal documents not intended for wide publication usually get produced in French, German and English, while communication between staff tends to be in French with English also widely used. The attitude to language is generally refreshingly pragmatic. One can expect officials, but not supporting grades, to be able to understand French and English and to be able to speak and write in at least one of these languages.

The Commission has a reputation for being slow and bureaucratic. While true that there is a fairly rigid hierarchy and set of procedures, the majority of the staff are very able and highly committed. In general, the services are understaffed but the process of developing proposals is also slowed by the wide ranging consultations undertaken and the very elaborate legislative process in which the Commission plays a continuous role. Translation of documents is also time consuming.

The organisational rigidity means some parts of the Commission may have time on their hands while others are working long hours. Organisational support is poor and adequate resources, taken for granted in most organisations, such as stationery or computers and software can be hard to come by. Lack of hard information about markets, industries or even legislation in the Member States is sometimes a problem. Consultants are frequently engaged to carry out studies, many of which are less informative than expected. Personal contact is not difficult once the most appropriate part of the Commission and direct phone number is established. However, staff are generally hard pressed and prefer contacts through European representative bodies, such as European trade associations.

A cheap and sometimes effective way of dealing with a problem involving a breach of Community rules can be by making a complaint to the Commission. David Medhurst, in his book *A Brief and Practical Guide to EC Law* quotes the case involving Brass Band Instruments Ltd who were being denied parts by Boosey and Hawkes. The latter enjoyed a dominant market position which was allegedly being 'abused'. Brass Band Instruments complained to the Commission who adopted an interim decision requiring Boosey and Hawkes to supply the part or face a penalty of 1000 ECUs per day. Another example of a complaint to the Commission, this time involving a breach of the rules on public procurement in relation to a contract on the Danish Storebælt bridge, is quoted in chapter 6.

Most of the Commission is divided into 23 Directorates Générales (DGs) each of which is the responsibility of a Commissioner. The following DGs may at times be of interest to companies:

DGI	External relations; conducts negotiations on the GATT agreement on behalf of the Member States, etc.
DGIII	Internal market and industrial affairs; deals with standards, public procurement, services, intellectual property, amongst others.
DGIV	Competition; deals with abuse of dominant positions and other anti-competitive company behaviour, state aid to industries, mergers, etc.
DGV	Employment, social affairs and education; deals with mutual recognition of qualifications, the ERASMUS and COMETT programmes (European Community Action Scheme for the Mobility of University Students, and COMmunity programme in Education and Training for Technology); also the administration of the Social Fund (see chapter 12).
DGVI	Agriculture; responsible for disbursing enormous sums of money under the Common Agricultural Policy (CAP).
DGVII	Transport; is removing some of the barriers to competition in this protected sector; also promoting the development of a balanced transport infrastructure, partly by means of a fund intended to support large projects with a European dimension.
DGVIII	Development; administers the Lomé agreement funds for projects designed to help developing countries.
DGX	Environment, consumer protection and nuclear safety.

DGXII	Science research and development.
DGXIII	Telecommunications, information industries and innovation.
DGXV	Financial institutions and company law.
DGXVII	Energy.
DGXXI	Customs union and indirect taxation.
DGXXIII	Enterprise policy, distributive trades, tourism and co-operatives; concerned with schemes designed to aid small and medium sized enterprises; has created a network of centres to provide advice and information around the Community – the Euro Info Centres (often known as *Euroguichets*, as in French).

The European Parliament

The European Parliament is based for the moment in Strasbourg, in the Alsace region of France. It consists of 518 members who are directly elected, from large constituencies in each Member State. Every state except the UK uses some form of proportional representation for their election. It struggles with considerable practical problems since most of the Committee sessions are held at its premises in Brussels about one kilometre from the Berlaymont, while many of its administrative facilities are housed in Luxembourg and its plenary sessions are held in Strasbourg. A new building is being built in Brussels which conveniently happens to be suitable for the plenary sessions. Many, including most Members of the European Parliament (MEPs), feel that all of its activities should be moved to Brussels, while the French vehemently want it kept in Strasbourg.

The Parliament's main role is to give opinions to the Council before both the 'common position' and final adoption stages of the legislative process. However, these opinions are sometimes treated casually, and the MEPs are anxious to have more power in the legislative process, and to be given the opportunity to propose measures themselves. The other Community institutions, and most of the national parliaments, tend to view the European Parliament, and the stream of resolutions it produces, as naive or simply barmy. The Parliament has two weapons which, because of their extreme nature, are not of much value. The first is that it can sack all 17 members of the Commission, and the second, which has once been used, is that it has the power to withhold approval of the Community budget. Despite its relative lack of power, a certain amount of lobbying has been focussed on MEPs, for example in connection with the proposed levy on blank cassette tapes to be used to reimburse performers who lose income through illegal copying. The recent directive on software copyright also led to intensive lobbying by rival groups of manufacturers.

MEPs can sometimes be of direct help. They are entitled to ask questions of the Commission, which must be answered. One answer sheds a fascinating light on the UK's phobia about rabies, one of the arguments put forward against the construction of the Channel Tunnel. Mr James Ford asked the Commission, in written question No 2161/88 of 6 February 1989, to specify the number of rabies deaths in each of 'the Twelve' for the last five years. The Commission answer, on 28 April, was: one death in Belgium in 1981, two deaths in France (1984 and 1986), one death in Germany (west), and *three in the UK*. Admittedly the three certainly became infected elsewhere, but the problem is small, and dwarfed by other causes of death such as road accidents.

The Court of Justice of the European Communities

The Court of Justice, based in Luxembourg, consists of 13 judges, including one from each Member State. It is responsible for interpretation and application of Community law, and until recently was the 'industrial tribunal' for disputes between Community institutions and their staff. The judgments of the Court of Justice are binding on Member States and take precedence over national law. The relationship between national law and Community law is discussed further in chapter 4. The Court is supported by the recently established Court of First Instance, also in Luxembourg, which deals with various matters, including disputes between the Community institutions and their staff.

Other institutions

Economic and Social Committee

Economic and Social Committee, usually referred to as ECOSOC, has 189 members representing employers, trades unions and various specialist areas appointed by the Council from lists of possible candidates supplied by the Member States. It is based in Brussels. The Council must consult the Committee as part of the legislative process (described in chapter 4). An article in the *Financial Times* (6 February 1991) about the ECOSOC was entitled 'The ultimate EC committee:

unheard of and unheeded'. Nevertheless, its balanced and considered opinions on legislative proposals are useful to the Commission, whose staff attend ECOSOC sessions to argue their case.

Court of Auditors

The Court of Auditors, based in Luxembourg monitors the Community's revenue and expenditure. It has of late submitted some very critical reports about procedures and fraud.

European Investment Bank

The European Investment Bank (EIB) operates from Luxembourg and finances capital investment projects which contribute to the development of the Community. (The activities of the EIB are referred to in chapter 12.)

European Bank for Reconstruction and Development

The European Bank for Reconstruction and Development (EBRD, though commonly known as BERD from the French arrangement of its name) was recently created to provide loans for the reconstruction and development of eastern Europe and thereby promote 'the fundamental principles of multi-party democracy, the rule of law, respect for human rights and market economics'. It is based in London and was set up with capitalisation of 10 billion ECU. (The activities of the EBRD are referred to in chapter 12.)

Chapter 3

The Community Construction Industry

There is no Community construction industry as yet: it remains 12 separate, and very individual industries operating in widely varying economic and physical climates, to different standards and regulations, under different forms of contract, different organisational arrangements and with differing liabilities. While the industry in Spain is busy installing new infrastructure and growing strongly, in the UK there is less infrastructure construction, and about 40 per cent of the output of the construction industry is on repair and maintenance (41 per cent in 1989).

The variety means that aggregate figures for the EC mean little. Much more interesting are comparisons between Member States and with other developed countries such as Japan and the USA. However, the industry contributes about 7 per cent of value added in the Community, and employs a similar proportion of the overall workforce. The industry is important as a source of employment for semi-skilled and unskilled labour. It creates the built environment, the transport, and energy supply networks and telecommunications facilities associated with economic development, the offices, shops and factories in which people work, and the water supply, sewerage disposal and sea defence works needed for health and safety. Its products are large, costly, usually visible, and if they fail widespread damage and disruption can occur. Construction is important in any developed and dynamic society.

Structure of the construction industry

The construction industry in most countries consists of a large number of very small companies and relatively few large companies. In the UK and French construction industries the large companies are larger, and carry out a greater proportion of the overall work, than in the countries such as Germany or Italy where regions are more

Table 3.1 Gross Domestic Product and Gross Domestic Product attributable to housing and to non-residential building and civil engineering

	GDP 1988	Per cent of 1987 GDP in:		1988 GDP in:	
		housing	non-res. bldg & cvl engineering	housing	non-res. bldg & cvl engineering
	ECU billion	%	%	ECU bn	ECU bn
W Germany	1017.6	4.88	6.39	49.61	65.00
France	804.2	5.37	6.13	43.16	49.30
Italy	701.8	5.19	4.66	36.45	32.68
UK	696.6	4.13	4.70	28.76	32.77
Spain	287.9	4.50	8.66	12.96	24.94
Netherlands	193.2	5.50	5.37	10.63	10.38
Belgium	127.0	3.19	6.03	4.05	7.66
Denmark	91.0	4.23	6.19	3.85	5.63
Greece	44.4	4.59	5.74	2.04	2.55
Portugal	35.3	5.31	8.52	1.87	3.01
Ireland	27.5	3.71	4.37	1.02	1.20
Luxembourg	5.6	4.14	8.64	0.23	0.48
USA	4076.0	4.96	7.49	202.13	305.28
EC 12*	**4032.2**	**4.86**	**5.86**	**195.96**	**236.29**
Japan	2403.6	6.13	11.56	147.35	277.84
Sweden	153.8	4.45	5.79	6.85	8.91
Austria	107.7	4.84	7.24	5.21	7.80

Notes: (1) The values in the final two columns have been obtained from the 1988 GDP assuming the percentages were unchanged from 1987.
(2) The figure for EC 12* excludes eastern Germany.
Source: Eurostat (partly derived)

discrete. Tables 3.2 and 3.3 demonstrate the relative importance of the large construction companies in the UK and France.

Amongst large companies, therefore, UK and French construction firms play a dominant role. Two thirds of the turnover of large construction companies in the UK, France, Germany, Italy and Belgium is accounted for by the large French and British construction companies. One factor behind the importance of large companies is the way the industry is organised. In Germany much work is done by trade contractors without a general contractor, whereas in the UK and in France to some extent the tradition has been for general contractors to be employed. However, in the UK since 1983 (date of the data in the tables), the number of small firms has grown substantially, reflecting

Table 3.2 Turnover of construction companies with 500 or more employees as a percentage of turnover of all companies employing 20 or more

	%
UK	42
France	37
West Germany	24
Belgium	17
Italy	17

Source: Eurostat, 1983

Table 3.3 Combined turnover of construction companies in certain Member States employing 500 or more, and as a percentage of all such companies

	ECU million	%
UK	16456	36.0
France	13968	30.5
Germany	11174	24.4
Italy	3392	7.4
Belgium	736	1.6
TOTAL	45726	100.00

Source: Eurostat, 1983

an increase in the growth of sub-contracting, as well as a shift towards practices such as management contracting and construction management which are similar to the German system.

Labour and productivity

The number of persons working in building and civil engineering in the Community (including self-employed and apprentices) was about 8.8 million in 1985 representing 7.2 per cent of all employment (see Table 3.4). This has probably increased slightly since. The number of employees (wage and salary earners) was recorded as 6.9 million in 1985.

There are large variations in income and hours worked in the industry (see Tables 3.5 and 3.6) (the wage data are for industry generally, but are indicative for the construction industry).

Table 3.4 Population, employment in industry and in building and civil engineering (1988)

	Population 1000s	Employment in industry NACE 1–5 1000s	Employment in building and civil engineering NACE 5 1000s
W Germany	61450	9771.0	1427.6
France	55884	5838.8	1207.0
Italy	57452	—	—
UK	57065	6751.4	1043.5
Spain	38765	3220.4	747.9
Netherlands	14760	1345.0	315.6
Belgium	9902	928.5	164.7
Denmark	5130	685.7	167.7
Greece	10004	937.6	232.1
Portugal	10287	—	—
Ireland	3538	300.0	62.6
Luxembourg	375	52.7	15.2

Note: NACE is the Community nomenclature for industrial activities.
Source: Eurostat

Table 3.5 Average gross hourly earnings of manual workers in industry

	Local currency/hr 1988	ECU/hour 1988
Denmark	87.62	11.02
W Germany	18.53	8.93
Luxembourg	338.0	7.78
Netherlands	17.99	7.70
Belgium	311.0	7.16
UK	4.53	6.82
Ireland	4.94	6.37
Spain	660.0	4.80
Greece	488.0	2.91
Portugal	288.0	1.69

Source: Eurostat

In addition to the above, non-payroll costs vary considerably from a maximum of 93.7 per cent of the direct wages for Italy, 84.9 per cent for France, 67.8 per cent for Germany, 33.8 per cent for the UK, down to 19.6 per cent for Denmark. In Denmark, the lavish social security system is funded from income tax, VAT and other forms of taxation (all figures for 1984) rather than from employees' and employers' contributions.

Employee productivity, measured by value of output per head, is also very varied (see Table 3.7), with the UK faring particularly badly.

Table 3.6 Annual working hours in the construction industry proper (person-hours worked, 1984)

	Manual and non-manual
Ireland	2038
Portugal	1990
UK	1953
Denmark	1728
Italy	1718
France	1698
Luxembourg	1681
Netherlands	1647
Germany	1635
Belgium	1347

Source: Eurostat

Table 3.7 Construction output per head and compensation per employee (1987)

	Output per head Index, UK = 100	Compensation per Employee Index, UK = 100
USA	156	162
Japan	143	151
France	132	—
Italy	113	103
UK	100	100

Source: OECD National Accounts

Four groups of Member States

It is possible to identify similarities between the industries in four groups of countries.

(1) *The UK and Ireland* share similar legal traditions and similar

construction industries. Referring particularly to the UK there are a number of characteristics. The tradition is to use an independent professional, usually an architect for building and an engineer for civil engineering, who carries out design and contract management; architects carry out periodic inspections and issue instructions on behalf of the client, whereas engineers normally closely monitor methods, work and progress. A feature of the role of the architect or engineer is that, although his services are paid for by the client, he is expected to fulfil a quasi-judicial role in interpreting the contract without prejudice towards either party. In the traditional system a general contractor, normally selected by the restricted procedure (called single stage selective tendering), is not involved in any way with the design of the permanent works. Increasingly the general contractor is using sub-contract trade labour to carry out work on site, and himself fulfils mainly a management role.

The industry is also characterised by a great variety of professionals, who collectively cost rather more than in other countries. These include the architect (this is the only professional title in the industry which is protected by law) whose education is less technical than in other Community countries, the quantity surveyor, who has no direct parallel in these countries and is concerned primarily with cost and contractual matters as the architect may direct, and a variety of engineers (structural, mechanical, electrical, etc.). Regulation is at a minimum, being primarily concerned with minimum standards for health and safety and heat loss of buildings. There is no restriction on fulfilling any of the roles in the industry, whether professional or trade (with a few very specialised exceptions such as the statutory inspection of dams). Another feature of the industry, associated with the sharp divide between design and construction, is the emphasis given to the detailed wording of the contract. This is partly a function of the English and Scottish legal traditions and partly a reflection of a society which, following the American model, is becoming more litigious. Clients tend to put emphasis on price and speed, often at the expense of quality, both of materials incorporated and of workmanship.

(2) *Denmark, Germany and the Netherlands'* systems share similarities and could be said to follow a northern European model. This is highly regulated, has formal procedures, and requires high standards. Unlike the first model, issues are dealt with on a personal basis by the parties involved rather than by reference to the contract. Developing a long-term business relation is paramount. From a British point of view, the business environment in these Member States seems clear and rational. However, the tendency for the state to regulate quali-

fications, contracts, remuneration for professionals and labour, working hours, and of what can do what seems alien, not to say intrusive. The emphasis on quality is perhaps the key.

(3) *Belgium, Luxembourg and France* share what can be termed the French model. Like the northern European model, the French one is characterised by being highly regulated and using formal procedures. Otherwise, the French approach allows a variable standard of work: the level of quality depends on the project. Statutory and legal requirements are more complicated than those in the British model. Public procurement procedures are elaborately defined by law. A 10-year no-fault insurance covering the building is usually required, and is compulsory in France itself. The client's insurer can exercise much technical control over the whole construction process, and in France results in a system of pre-qualification for contractors which, though private, presents a hurdle for foreign contractors. Quality requirements and risk sharing are variable though clearly defined. Claims and cost overruns in excess of 20 per cent are normal. Politics, regulation and business become intertwined in the French system, in a way that the British may regard as corrupt, and certainly discriminatory.

(4) *Italy, Spain, Portugal and Greece* to some extent share a Mediterranean model. Quality is less emphasised than in the northern European model, but may be good, particularly in Spain and Italy. Public administration is slow to respond. Public procurement procedures are costly, and highly regulated by law. Private sector behaviour is governed by a network of personal and company relationships, including political; politics and business are closely related. The system appears bureaucratic and highly political. Planning, procedures and regulations can be labyrinthine and call for flexibility and creativity in dealing with them. Outsiders can have problems in adapting. Economic conditions are generally volatile.

Three procurement systems

Broadly there are three procurement procedures used for civil engineering and building in Europe. These are:

(1) General or lump sum contracting whereby an architect or other designer produces a design and a main contractor takes responsibility for construction using sub-contractors in some degree. General or lump sum contracting is commonly used in Italy, Spain,

Belgium, Greece, Portugal, the Netherlands and the United Kingdom. The amount of detail design done by the contractor varies considerably. In the UK the contractor usually does no design, while in France and Spain the contractor may do most of it.

(2) Trade contracting, whereby trade contractors are managed, normally by the designer, or construction management agent. As with the system above, the architect or designer is employed to conceive a design. This method is used in Belgium, France, Germany and is gaining popularity in the United Kingdom.

(3) Turnkey contracting, or design and build, whereby the contractor supplies the relevant design services and carries out construction. This may be used in Germany and is commonly used in the United Kingdom (mainly for building).

In some countries another system is used in which the private sector is contracted to provide a facility for public use. This is called 'concession contracting' and is known in the UK as 'build-operate-transfer' (BOT). It involves the private sector in selecting and financing projects as well as designing, subject to performance criteria, building and operating a facility for the public use. The contractor may be paid under some negotiated or competitively tendered arrangement, or more usually the concession contractor is given the right to charge the users of the facility. A mixture of the two can also be used. For example, for the Manchester Metrolink light rail scheme a central government grant was available and bidders were invited to tender for the contract, making due allowance for the income that they would be able to derive from ticket income and other sources during the 15-year period before the end of the contract. The construction of the Channel Tunnel is by Eurotunnel as the concession contractor, with in this case a 55-year period from the signing of the contract (allowing about 47 years of income generation before contract expiry). The system is widely used in France, for example for tolled motorway construction and operation, and in Italy, where a form of concession is also used for construction management. The latter may in some circumstances be a device for circumventing the local public procurement procedures.

Other aspects of national practice

General contracting is making inroads into the French market, but the most widespread system is still to employ trade contractors with contract management by the architect or by a member of the *bureau d'études techniques* (the nearest French equivalent to a British consulting engineer). In France contractors generally produce technical drawings, for example for reinforced concrete details. Trade contractors may sub-contract parts of their work although clients tend to dislike the practice. Where a lead contractor is needed, it is often the structural frame contractor who takes the role. In the UK it is regarded as bad practice, even unethical, to haggle with the contractor about the tender price; in France, however, trade contractors are expected to reduce their prices during post-tender negotiations, and make allowance for this when tendering.

In Germany, as in France, supervised trade contracting is the usual method of construction. Either the design architect will carry out project management, or a separate management firm will be appointed.

Even though general contracting is usual in the Netherlands, clients like to enter into contracts directly with the the piling, demolition and mechanical and electrical contractors; the ground conditions often mean that deep foundations are needed. Supervision and co-ordination of these direct contracts is usually done by the main contractor, who is paid a fee.

The contract arrangements in Belgium are largely similar to those in the Netherlands.

General contracts with contractors carrying out most of the work themselves using craftsmen and sub-letting building service installation are usual in Italy. The architect will usually carry out detail design. The situation is the same in Spain, except that the contractor usually carries out detail design and tends to employ trades sub-contractors. An unusual feature of Spain is that the client pays fees for the architect's services to the local 'college of architects' which then pays the architect; the whole system is highly regulated with the architect's role enjoying legal protection. There are also 'technical architects' whose role has similarities with those of the resident engineer and quantity surveyor in the UK.

In Portugal foreign developers and contractors (mainly British) are dominant. The role of the architect usually ends with planning permission. Both architects and engineers must be registered to practise. The role of the engineer, who carries out site supervision, is more powerful than in Spain. Contractors usually carry out detail design. Design and build has also had some success in Portugal. In general the system is less regulated than Spain's, but changes are taking place, at least partly in response to pressures associated with membership of the Community.

In Greece, a few major construction companies work as general contractors in association with large numbers of unregistered sub-contractors, many of whom operate on a labour-only basis.

Convergence

There are major differences in many aspects of the construction markets and the way work is carried out in the Member States, but the single market is bringing about important changes which will tend to bring the economic and the operating conditions closer. However, more than anything the internal market has stimulated the national industries to become aware of the potential of the markets in other Member States, and to recognise that differences can be exploited and that there are important lessons to be learned. The interest and cross-border activity, particularly in the form of takeovers and joint ventures, is likely to cross-fertilise the national industries and through a form of evolution allow the most fit procedures, practices and technologies to emerge. The next chapter examines briefly the framework of Community law and the legislative process. Following chapters examine the key areas of the internal market programme. A series of country studies (chapter 13) then outline the cultures, markets and construction industries in each of the Member States.

Chapter 4

Introduction to European Community Law

European Community law has brought profound changes into national law, including British, to the dismay of some. National legislation deriving from Community law must be interpreted using the principles of Community law, which are fundamentally different from those in British law. British law is interpreted literally, while in Community law the purpose must be considered in interpretation. To UK eyes the wording of Community law may appear general, but since the purpose and background are spelled out in the preambles, the end result may be more certain. Even the preparatory documents may be interpreted to shed light on the purpose. British lawyers are used to minute analysis of wording in search of loopholes. Those drafting bills or contracts respond by lengthy and heavily qualified texts. Using the purposive approach this might not be necessary. In *Litster* v. *Forth Dry Dock Co Ltd* (see p 53) the strict interpretation of the wording of a statute implementing an EEC directive suggested a way of circumventing the regulation. However, when purpose was considered the apparent loophole was held not to exist.

A further, important, rule is that Community law has primacy over national law. National law must be amended to be compatible with the EEC treaty or directives introduced under it. The intention of directives appears to be upheld even if incorrectly transposed into national law, or even if not transposed at all. This is important, because some Member States, Italy for example, have a very poor record of implementation. Moreover, where national law and Community law conflict, national courts do not even have to wait for the national law to be changed (e.g. by an Act of Parliament). If the Community law is 'directly effective' national courts are obliged to give effect to it in any case before them, and to ignore the inconsistent provisions of national law. A recent ruling by the Court of Justice took this a stage further. The court held that judges in Member States also have the power to withhold the application of a national law pending a decision of the European Court about its compatibility with EC law (see the *Spanish fishermen* case p 56).

The influence of Community law is being felt increasingly in the courts. It represents a new way of thinking and its effects are likely to permeate throughout legal systems in the Community. A new convergence in legal doctrine and even legal systems could emerge.

The influence of the treaties is probably strongest in the business sphere, but it affects employment and environmental legislation. Legislation governing competition between companies is central; indirect taxation, for example whether Value Added Tax should be charged on new building, is affected, as are a wide variety of areas including health and safety at work, consumer protection, procurement procedures, even for some private companies, and many others.

Community developments affect domestic issues as well as international ones. There are sometimes new rights and remedies not available in national law. For example, under the new compliance directive dealing with failure to comply with the public procurement directives (see chapter 6) it may be possible to claim damages if a public authority rejects a tenderer on inadmissable grounds.

Community law thus is crucial even to companies and to individuals with no intention of working or trading abroad. It is integral to the body of law which affects every company and citizen throughout the Community.

The Treaties

The three treaties setting up the European Communities are:

(1) The European Coal and Steel Community Treaty, known as the Treaty of Paris, which was signed on 18 April 1951 and came into force in July 1952. It will expire in 2002.
(2) The European Economic Community Treaty, referred to as the Treaty of Rome, signed on 25 March 1957 and which came into force on 1 January 1958.
(3) The European Atomic Energy Community Treaty which was also signed in Rome on 25 March 1957 and came into force on 1 January 1958. It is usually known as the Euratom Treaty.

In addition there are supplementary treaties dealing with the accession of the UK, Ireland, Denmark (signed 22 January 1972, in force from 1 July 1973), Greece (signed 28 May 1979, in force from 1 January 1981) and Spain and Portugal (signed 12 January 1986, in force from 1 July 1986). The provisions allowing free movement of labour in relation to Spain and Portugal will only come into effect on 1 January 1993.

The Single European Act

The Single European Act (signed in Luxembourg 17 February 1986, in force from 1 July 1987) made a number of important changes to the original EC treaties. It is also known for having set the target date of 31 December 1992 for the completion of the Single European market. By introducing a form of majority voting it removed a key obstacle to its creation. Amongst other changes it also introduced new articles into the EEC treaty on economic and social cohesion, research and technological development and the environment.

The EEC treaty is the most significant for industry in general. In this book 'the Treaty' is used to refer to the EEC Treaty except where the context indicates otherwise. An edition of the treaties, the Single European Act and all significant protocols, annexes and associated text, is available from the Office of Official Publications of the European Communities in Luxembourg, or HMSO [4.1]. For ease of use all the changes and additions introduced by the Single European Act are consolidated into the text.

Application of Community law in the UK

The law of the European Communities was made applicable in the UK by means of the European Communities Act 1972, section 2(1) of which reads:

> 'All such rights, powers, liabilities, obligations and restrictions from time to time created or arising by or under the Treaties, and all such remedies and procedures from time to time provided for by or under the Treaties, as in accordance with the Treaties are without further enactment to be given legal effect or used in the United Kingdom shall be recognized and available in law, and be enforced, allowed and followed accordingly.'

Some articles of the treaties do not apply directly, and need measures, possibly a directive for example, to have any effect. Other articles need no implementation: these are said to be 'directly effective'. An example of the latter is Article 30 of the EEC Treaty which prohibits barriers to the free movement of goods:

> 'Quantitative restrictions on imports and all measures having equivalent effect shall, without prejudice to the following provisions, be prohibited between Member States.'

In a case involving this article an Irish authority had insisted on using only pipes which had obtained a test certificate to an Irish standard. The court held in *European Communities* v. *Ireland* [4.2] that Ireland had failed in its obligations under this article.

Directives, regulations and decisions

The EEC Treaty and the ECSC Treaty give different meanings to certain words, and this can confuse! An ECSC recommendation is equivalent to an EEC directive and an ECSC general decision is equivalent to an EEC regulation. The EEC meanings will be used in this book.

Regulations

Regulations are adopted by the Council or the Commission and are binding and directly applicable in every Member State without any further enactment being needed. Regulations tend to be more technical and less far reaching than directives. Examples include Council Regulation 2137/85 [4.3] which introduced a new form of corporate entity (the European Economic Interest Grouping) designed to make cross-border collaboration and joint ventures more straightforward (see chapter 5 for details); Council regulation (EEC) 17 of 6/2/62 [4.4] introduced a system for companies to apply to the Commission for approval to enter into agreements which would otherwise be prohibited under treaty articles dealing with competition. Many regulations deal with agriculture.

Directives

An EEC directive is an instruction addressed to the Member States to achieve a given legislative result by a given deadline. This is usually done by transposing the terms of the directive into national legislation. Thus the public works directive 71/305/EEC [4.5] (see chapter 6) is an instruction to the Member States to modify their public procurement procedures to comply with the requirements of the directive. In France this was done by amending the *code des marchés publics*, a code which is part of national law. In the UK there is no equivalent body of law governing public procurement, but the government has sought to achieve compliance by issuing instructions to all public purchasing authorities within the scope of the directive in the form of circulars.

Whether this approach fully achieves the results required is not certain. An example of a directive implemented in the UK by means of a statute is Council Directive 85/374/EEC [4.6] dealing with liability for defective products. The Consumer Protection Act 1987 implements this directive in the UK and makes specific reference to it.

Most important Community legislation is in the form of directives. Since they are intended to be given effect by being transposed into national law one might suppose that the directive itself would have no effect unless transposed. However, a number of more recent cases have shown that the EC directives can have direct effect, for example if the directive has not been transposed into national law by the due date or where incorrectly transposed. In *Pubblico Ministero* v. *Ratti* [4.7] a directive concerned with labelling of chemicals was held to be of direct effect even though not implemented at the time (remember Italy's poor record for implementation).

In *Marshall* v. *Southampton and South West Hampshire Area Health Authority* [4.8] Miss Marshall successfully appealed against a requirement that she should retire at 60 instead of at 65 contrary to a directive which provided for equal treatment between men and women. The European Court stated that:

> 'Wherever the provisions of a directive appear, as far as their subject matter is concerned to be sufficiently precise, those provisions may be relied on by an individual against the state, where the state fails to implement the directive in national law by the end of the period prescribed or where it fails to implement the directive correctly.'

Decisions

Decisions are addressed to specific persons or organisations and are immediately binding on those to whom the decision is addressed. No implementation into national law is necessary. Many Commission decisions are issued in connection with competition cases. For example, in Commission Decision 86/138/EEC Eurofix/Hilti [4.9] the Commission ruled that Hilti's practice of tying the sale of nails to its patented cartridges was an abuse of its dominant position in the market for nail guns and contrary to Article 86 of the Treaty. Commission Decision 88/568/EEC [4.10] stated that it had no grounds for action under 85 of the Treaty (which is concerned with undertakings combining against the interest of competition) in connection with the construction contract and the *maître d'oeuvre* contract entered into by the Channel Tunnel Group Ltd, France Manche SA and Eurotun-

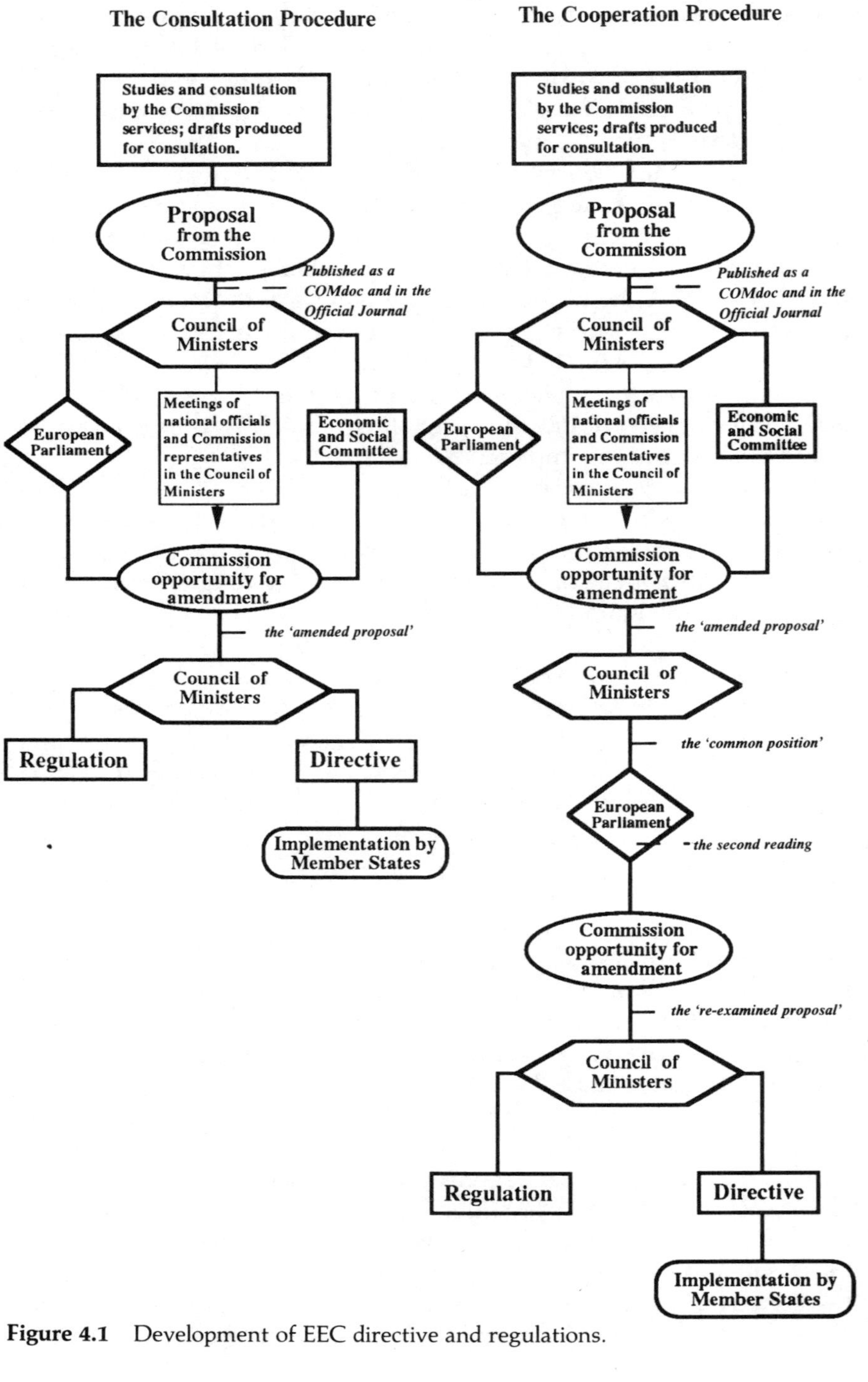

Figure 4.1 Development of EEC directive and regulations.

nel. A group of five of the largest UK contractors and five of the largest French contractors could in other circumstances act together in a most anti-competitive way!

Recommendations and opinions

As the names suggest, recommendations and opinions are not binding in law, though they may have some influence on the court. A 'reasoned opinion' or *'avis motivé'* is used as the first step in any infringement procedure taken against a Member State, for example where the requirements of the public works directive appear not to have been complied with.

Article 169 (EEC Treaty):

> 'If the Commission considers that a Member State has failed to fulfil an obligation under this Treaty, it shall deliver a reasoned opinion on the matter after giving the State concerned the opportunity to submit its observations. If the State concerned does not comply with the opinion within the period laid down by the Commission, the latter may bring the matter before the Court of Justice.'

The consultation procedure for adoption of directives and regulations

This procedure is required for legislation under Article 100 of the Treaty and was the main procedure available before the Single European Act. It requires unanimous voting in the Council. See Figure 4.1 for a flowchart depicting both the consultation procedure and the co-operation procedure.

The co-operation procedure for adoption of directives and regulations

This procedure was introduced by the Single European Act and is required for legislation under Article 100a of the Treaty. Because it introduced qualified majority voting instead of unanimous voting in the Council it unblocked the legislative process and made the completion of the internal market within a finite timescale feasible. It can be used for 'measures for the approximation of the provisions laid down by law, regulation or administrative action in Member States which have as their object the establishment and functioning of the internal market'. The qualified majority voting procedure cannot be used for 'fiscal provisions, to those relating to the free movement of persons [or] to those relating to the rights and interests of employed persons'. Article 149 describes the details of the co-operation procedure. Refer to Figure 4.1 for a flowchart of the procedure.

Proposal from the Commission

The first stage in the procedure is a proposal from the Commission. Much study and consultation precedes its publication. Studies may be commissioned, committees and trade bodies consulted. Special government and industry consultative committees have been established by the Commission in most sectors and receive a regular supply of working papers to be discussed in periodic meetings. Before the proposal is formally adopted all parts of the Commission with a possible interest are consulted. DGXXIII (a Commission Directorate Général) vets all draft proposals for their impact on small and medium enterprises.

Language differences slow up the process, since the document must be available in all nine official Community languages. There is a group of lawyers called the *juristes linguistes* who carry out the final check. The Commissioners themselves consider proposals in weekly formal meetings. After adoption by the Commission proposals are transmitted to the Council.

Proposals are issued with a reference such as COM(89) 141, and are one type of document known as the 'COMdocs'. They eventually get published in the *Official Journal*, but can usually be obtained by contacting the appropriate part of the Commission or from the national department responsible for consultation – for example the UK Department of Trade and Industry.

Action following transmission of the proposal to the Council of Ministers

The Council issues the proposal to the European Parliament and to the Economic and Social Committee for their comments and proposed modifications. The Council also sets up a committee of national civil servants with expertise in the field of the proposal. The national ministries embark on consultations with industry and other departments. In the Council working group meetings a detailed analysis is conducted, with Commission staff in attendance to argue their case and learn about possible objections.

Economic and Social Committee and European Parliament consultation

The Economic and Social Committee conducts a detailed study of the proposal, with Commission representatives at hand to give reasons and reactions to the comments being made. A parliamentary committee chaired by a 'rapporteur', with Commission staff as ever in attendance, considers the proposal in rooms in the Parliament's building in Brussels, 10 minutes walk from the Berlaymont.

The Commission is free to modify its proposal at all stages during the entire process. The qualified majority voting procedure can only be used if the Council is acting on a Commission proposal. The Council can only change a Commission proposal by unanimity. The Parliament carries some weight insofar as its opinion is indispensible to the adoption of a proposal. Amendments adopted by the Parliament and supported by the Commission stand a good chance of being adopted by the Council. Commission staff are therefore likely to spend some time negotiating amendments with the rapporteur.

The plenary session of Parliament (sitting for the moment in Strasbourg) votes on amendments tabled by the specialist committee. The Commissioner responsible for the proposal normally goes to Strasbourg to attend the plenary session armed with briefing notes about details of the proposal, and accompanied by Commission staff who have worked on it. A badly briefed or inattentive Commissioner may miss the finer details and could make serious mistakes during the session. The hierarchical nature of the Commission requires that the level of the representative is closely matched to the level of the occasion. Given the political implications this is not surprising, but the expert member of the Commission staff may be reduced to murmured explanations in the meeting and signalled warnings to the boss if a blunder seems imminent.

The amended proposal

Once the opinions and advice have been received from the Economic and Social Committee and Parliament, the Commission draws up an 'amended proposal'. This involves detailed discussion and careful drafting of amendments, which could number as many as a hundred, and often involves discrepancies amongst languages, differing views amongst Directorates Générales, hindrances such as inadequate computer and printing facilities and reliance on temporary secretarial staff. It also requires the same modifications to be made in nine languages, including a different alphabet (Greek) and a host of special accents and characters, not to mention nuances which are difficult to translate accurately between two languages let alone nine.

The Council working group then examine the amended proposal.

Once the Council thinks that the differences amongst the representatives of the Member States have narrowed down to a handful of key issues, discussion moves to the COREPER meetings (Comité des Répresentants Permanents), where political factors will play a more important part. Rapid progress is made in the COREPER meetings, partly through the pressure of a heavy workload and a wide range of technical matters to be dealt with. The Commission is constantly seeking to find the source of problems and by various informal and tactical means to manoeuvre the Member States to bargain amongst a multitude of issues. Meetings are chaired by the representative of whichever Member State holds the presidency of the Council at that particular time.

The common position

The objective is to reach a 'common position', a term used in Article 149(2)(b) of the Treaty. If possible the COREPER will settle every detail so that the national ministers can adopt the common position without discussion, but key issues often remain.

Council meetings at ministerial level are held periodically in Brussels or Luxembourg. The Commissioner responsible represents the Commission as in the plenary session of Parliament.

The common position is published in the *Official Journal*, and sent to Parliament with information about reasons which led the Council to adopt the common position and about the Commission's position. Parliament may approve the common position, or within three months it may 'by an absolute majority of its component members, propose amendments to the . . . common position' or reject it. Under the internal rules of the Parliament, discussion is restricted to matters raised earlier, but the wild and the woolly are included amongst parliamentarians and restraint may not be easy.

Re-examined proposal

The Commission includes as many of the crop of amendments voted for in the second reading of Parliament as it can. Unacceptable modifications may be undermined, possibly by using 'may' instead of 'shall'. Outright rejection could undermine co-operation and offend Members of Parliament. The Commission needs the support of the democratically elected body even though it can be a nuisance. One day the European Parliament may be given more power. The re-examined proposal, in nine languages, is delivered to the Council after adoption by the Commission.

Adoption

The procedure in Council for adoption of the directive follows the same pattern as for the adoption of the common position, with meetings of the experts, COREPER and finally the ministers themselves.

Directives

Strangely, under the treaty, directives do not have to be published, but are merely to be 'notified to those to whom they are addressed', i.e. to the Member States. This is because directives have to be implemented by Member States and, as explained above, it was thought that they could not have direct effect. In practice they are published, in the L series of the *Official Journal*, but usually mixed in with a variety of other measures such as agricultural regulations.

Preambles

After a long-winded title, not designed for convenient reference, come the recitals or preambles. The Treaty article(s) for the legal basis comes first. Wherever possible the base will include article 100a (the internal market article), 118a (health and safety) or other article allowing the co-operation procedure to be used. The Commission proposal for a directive on working hours, which so outraged the Federation of Civil Engineering Contractors and other UK employers organisations, was presented as a health and safety measure rather than an employment measure for this reason. Member States may challenge a measure with an unsound legal basis. Remaining preambles generally give the context for the directive and explain the purpose of the directive and its provisions, but in very general language. The Economic and Social Committee and Parliament opinions are also referred to. A preamble from the Council Directive of 25 July 1985 on the 'approximation of the laws, regulations and administrative provisions of the Member States concerning liability for defective products' 85/374/EEC [4.11] illustrates the generality:

> 'Whereas, to protect the physical well-being and property of the consumer, the defectiveness of the product should be determined by reference not to its fitness for use but to the lack of the safety (sic) which the public at large is entitled to expect; whereas the safety is assessed by excluding any misuse of the product not reasonable in the circumstances.'

The articles

The articles themselves are more simply written and with fewer qualifications than clauses in UK legislation. This, and unfamiliar terminology, may give the appearance of ambiguity and vagueness. However, the preparatory documents and the preambles can be used for interpretation. The court may even refer to the Commission to find out what was intended. Such useful things as indexes, tables of contents or headings (other than chapter headings) are not provided. But directives whose function is to amend previous directives bring the greatest difficulties in understanding. For example 89/440/EEC [4.12], brought major changes to the original public works directive 71/305/EEC [4.13] and in fact left little of the original intact. The amending directive, with its own series of articles, refers back to articles in the original directive with deletions, substitutions and new provisions. The Council has instituted a policy of issuing consolidating directives replacing all former texts, but this can bring complications. For example, one of the annexes in the public supplies directive (see chapter 6) uses an outdated classification system. The replacement system does not bring about exact correspondence and slight shifts of meaning would arise if it were used. Since the consolidation document will have to be a directive in its own right the full co-operation procedure is involved. Another problem is that the temptation to tamper with the wording is difficult to resist.

Final articles deal with administrative matters such as the deadline for implementation and extended deadlines for recent members of the Community or (more often) Member States with economic problems. Monitoring by the Commission is often specified. Member States are also required to send the Commission the texts of the measures used to implement the directive, so that compliance can be checked.

Implementation

Implementation rates vary greatly between Member States. In September 1989 the Commission issued a report on implementation of the single market measures showing the UK, France, Denmark and the Netherlands with the best record, and Italy and Greece lagging far behind. Of the 69 measures which should have been in force at that time only seven had been implemented in all 12 states. Spain and Portugal, as more recent members were understandably struggling to accommodate more than 1000 directives already in existence when they joined, as well as trying to implement the flood of internal market measures. Even so, Italy had the worst record. Only 29 of the measures had been implemented, 23 had not been implemented and there had been 10 infringement rulings in the Court of Justice concerning failure to implement directives.

Courts of the European Communities

The Court of Justice of the European Communities is the primary court, but in September 1989 the European Court of First Instance was set up as provided for by the Single European Act. The new court will deal with employment disputes involving staff of the Community institutions, an industrial tribunal function which mixed awkwardly with the other functions of the Court of Justice. It will also deal with competition cases, which tend to be complicated and lengthy, and involve voluminous documents and intricate economic arguments. The new court was partly introduced to reduce the two-year queue of cases in the Court of Justice; however, the number of cases appears to be increasing exponentially.

The main functions of the Court of Justice are the interpretation of the Treaty and subsidiary legislation, and the judicial review of the acts of the Community institutions. Procedure is unlike that in the UK courts, rather less formal and with more reliance on written presentations. A *juge rapporteur* is appointed by the President of the Court to see the case through and to provide the court with a briefing on the facts of the case. Another feature of the court is the Advocate General who makes a reasoned and impartial assessment of the case. This will be in the form of a report to the court before the case is heard. The Advocate General then sits with the judges, but to one side, and after the case delivers his opinion. If the court agrees with the opinion it has the same authority as the judgment itself. Reports, which include the opinion of the Advocate General and the judgment, appear at infrequent intervals in the European Court Reports series. Some court reports, along with legal comment and analysis, are published in the Common Market Law Reports series. These are not official, but generally appear more quickly than the European Court Reports.

Who uses the European Court?

The Commission may act against Member States under article 169 of the Treaty. Member States may act against each other under article 170 of the Treaty. Another source of cases is when national courts refer points of Community law to the court. Occasionally individuals or companies can take direct action in the European Court, but only against a Community institution and provided it has acted so as to cause a direct effect on that individual or company. In general, if Community law relates to an individual or company, action would be taken in national courts with a possible request for reference to the European Court.

Under Article 173 'any natural or legal person may . . . institute proceedings against a decision addressed to that person or against a decision which, although in the form of a regulation or a decision addressed to another person, is of direct and individual concern to the former'. The Court of Justice is obliged to review the legality of acts of the Council and Commission, other than recommendations or opinions, on the grounds of lack of competence, infringement of an essential procedural requirement, infringement of the Treaty or of any rule of law relating to its application, or misuse of powers.

Article 173 specifies a two-month time limit for instituting the proceedings; if this limit is exceeded there is a way of getting individual relief, even though the regulation or other legislative act will otherwise remain in force. Under Article 184, '. . . any party may, in proceedings in which a regulation of the Council or the Commission is in issue, plead the grounds specified (described above) in order to invoke before the Court of Justice the inapplicability of that regulation'.

Interpretation of Community law

The Court uses the purposive (or teleological) approach to the interpretation of Community law. This approach is alien to UK law which traditionally favours the literal intepretation. In UK law the wording is paramount. One may be contrary to the spirit of the law but literal compliance is sufficient. The British approach leads to a microscopic reading of the text, a hanging on the precise meaning of each word, and if necessary a laboured search for loopholes. The legislators and authors of contracts and other legal documents need to anticipate this, but the result is often tortured and heavily qualified prose which defies understanding.

A different situation applies with purposive interpretation. If there is a gap in the main text reference may be made to the preambles. If the preambles fail to give guidance the Treaty, particularly the article or articles on which the measure is based, may aid interpretation. The plain meaning of the text is used as a starting point, but this may not necessarily be the same as the literal interpretation of the text. If there is a conflict between the wording of the article of a directive and the preambles, the preambles, which express the purpose, will take precedence. Even the context, for example the general purpose of that section of the document in which an article appears, may be used to aid interpretation.

An important consequence of the purposive approach was demonstrated in *Litster* v. *Forth Dry Dock Co Ltd* [4.14]. The Transfer of Undertakings (Protection of Employment) Regulations 1981 had been brought in to implement Directive 77/187/EEC [4.15] which was intended to give protection to workers when the ownership of an undertaking was transferred. The employer argued that the regulation only applied to workers who were actually employed at the time of the transfer. The worker in question had been dismissed, the regulations did not therefore apply, the employer argued, even if re-employment occurred within a few hours. However, in the House of Lords Lord Templeman stated that: 'The courts of the United Kingdom are under a duty to follow the practice of the European Court by giving a purposive construction to directives and to [national] regulations issued for the purpose of complying with directives.' It was held that the gap in employment at the time of transfer did not circumvent the regulation.

UK law implementing Community law cannot be interpreted in isolation; it must also be interpreted using principles adopted by the Court of Justice.

General principles

The Treaty and legislation made under it cannot be comprehensive. The Court of Justice has therefore had to develop general principles, which in practice have been derived from those used in the courts in the various Member States. These are set out by the court in giving explanations for their rulings. The principles generally mentioned are equality, legal certainty (which includes legitimate expectation) and general avoidance of retroactive effect. In relation to legal procedure the general principles include the right in natural justice of a hearing, the duty of authority to give reasons for decisions, and the right to challenge rulings.

Equality is embodied into the Treaty with regard to equal pay for equal work (Article 119), and within certain limitations to freedom from discrimination on the grounds of nationality, but it has been extended to a general principle of similar treatment for similar situations and to equality in the imposition of burdens by the law.

However the principle cannot be applied too blandly as pragmatic reasons have in the past been accepted for apparently discriminatory treatment. For example, British scientists and engineers on the Joint European Torus project in Culham, Oxfordshire, tried to invoke the principle to obtain equality of pay with colleagues from other Member States – *Alan Ainsworth and others* v. *Commission and Council of the European Communities* [4.16]. Under an agreement between the UK government and the Commission, British recruits to the project were obliged to join the UK Atomic Energy Authority while others joined as temporary servants of the Commission and enjoyed better pay and conditions. They failed on the ground that under the JET agreement the UKAEA was in a special position, bearing the cost of putting the site into 'standard condition' and providing scientific equipment as well as 222 staff out of a total of 384. The UKAEA had been anxious to prevent the situation from upsetting its current conditions and arrangements and had negotiated that staff made available to the project remain subject to its own conditions of service. The special circumstances were sufficient 'objective justification for the difference in treatment . . .'.

Importantly, the principle of freedom from discrimination on the grounds of nationality has been held not to be restricted to the action of public authorities, but to apply to rules of any other nature aimed at 'regulating in a collective manner gainful employment and the provision of services'. This was held in *Walrave and Koch* v. *Association Union Cycliste Internationale* [4.17] when Article 7 of the Treaty was used to challenge the rules of a cycling association.

The principle of legal certainty (which includes non-general retroactive application) were applied in the well-know Defrenne case – *Defrenne* v. *SA Belge de Navigation Aeriénne Sabena* [4.18]. It was held that Miss Defrenne, an air hostess who had been paid less than male cabin stewards for the same work, was able to claim back-pay for herself on the grounds of equality of treatment. However, others would only be able to claim back pay from the date of the judgment. The principle of legitimate expectation derives from German law. It was applied in *Mulder* v. *Minister van Landbouw en Visserij* [4.19] where a dairy farmer had been induced by a regulation to reduce milk production. He then found that under a new regulation introducing milk quotas that he could not obtain an adequate quota because the quota allocation was based on production in the year when it had been temporarily reduced. It was held that he could have legitimately expected to resume production in accordance with the rules of the earlier regulation designed to reduce production.

The Court will also adopt the principle of upholding fundamental human rights, which are not spelled out in the Treaty. In *Nold KG* v. *Commission* [4.20] the court stated that:

'In safeguarding [fundamental human rights] the court is bound to draw inspiration from constitutional traditions common to the Member States, and it cannot therefore uphold measures which are incompatible with fundamental rights recognised and protected by the constitutions of those States.

Similarly, international treaties for the protection of human rights on which the Member States have collaborated or of which they are signatories can supply guidelines which should be followed within the framework of Community law.'

Reference to European courts

For those other than the Commission and Member States direct action through the European Court is not usually possible. The way then is to use the national courts which may, if necessary, make a reference to the Court of Justice to obtain clarification and interpretation on Community law. When will requests for reference to the Court of Justice be granted? National courts may be unwilling to make a reference because of the delay and cost involved or because the issue appears to be so clear cut that a reference is unnecessary. However, the Court of Justice is anxious to avoid the problem of differences of interpretation growing up between Member States and unless the principles in question are so well established that the issue would be clear to both the European court and the courts of other Member States a reference should be made. The lower courts may simply avoid the issue by refusing to make a reference, leaving the decision to a higher court if an appeal is made. The problem of delay, which used to be about two years, may be reduced now that the Court of First Instance is in operation.

Remedies

It may be possible to obtain non-contractual damages from the Commission or other Community institutions that have exceeded their authority or where their improper action caused damage to private persons or companies, but this is unusual. In *Adams* v. *European Community Commission* [4.21], Mr Stanley Adams, an executive with the multinational pharmaceutical company Hoffman-La Roche, had become disillusioned with the practices of his company and had given evidence to the Commission alleging abuse of a dominant position in the market for vitamins. Unfortunately for Mr Adams the Commission allowed his identity to be discovered. This resulted in his arrest in Switzerland for the breach of that country's strict laws on business disclosure. He succeeded in obtaining damages for negligent breach of confidence by the Commission.

More likely would be action against a statutory authority or Member State for an injunction preventing it from acting in breach of its obligation under the Treaty or a directive. It might even be possible to obtain damages, though at present this is not certain.

In *Bourgoin SA* v. *Ministry of Agriculture* [4.22] some French turkey breeders sued the Minister of Agriculture in the UK for refusing to allow their produce to be imported, in breach of Article 30 of the Treaty, which prohibits barriers to the free movement of goods. The English court was divided about whether the Minister could be liable for the way in which he exercised his legislative powers, but the action was settled for £3.5 million before the European Court had the chance to express itself. For the moment the presumption must be that statutory authorities, ranging from government departments to local authorities and quangos can be held liable for damages.

Spanish fishermen and new power for national courts

One case which shook the legal profession and made constitutional history, illustrating the growing power of Community law and its relevance to private parties, is the obscurely named *R* v. *Secretary of State for Transport ex Parte Factortame Ltd* [4.23]. It involved trawlers under Spanish ownership registered in British ports. (Factortame is the off-the-shelf company bought by one of the trawler owners.) The vessels were registered in a British port to take advantage of the UK fishing quota ('quota hopping'). The Community, perhaps inconsistently, allocates quotas on a Member State basis. One trawler company was operating out of a British port with British crews, but others were simply registered in British ports, and had Spanish crews and sold their catches in Spanish ports. To prevent Spanish fishermen benefiting from UK quotas, the UK government brought in the Merchant Shipping Act 1988 which restricted registration partly by reference to nationality (rather more cleverly the French government, with similar objectives in mind, introduced a language test for qualification purposes). The Act stated that 75 per cent of directors and shareholders in companies operating fishing vessels in UK waters must be British.

On a reference, the Court of Justice held that the nationality requirement was incompatible with the Treaty, particularly the provisions on the freedom of establishment. However, bearing in mind the problems inherent in allocating fishing quotas to Member

States the courts did not rule against a requirement that trawler owners should live in the UK – an elegant compromise. Once again national law had to cede precedence to Community law, but even more interesting was the question of whether national courts could grant interim protection from the application of a national statute pending a decision from the Court of Justice.

The House of Lords had previously held that there was no such right, but in the context of the Factortame case the Court of Justice held that national courts did indeed have the power to suspend the operation of a statute. Never before have UK judges had the power to do more with statutes than interpret them.

This case could settle yet another point of principle concerning Community law – the question of the liability of statutory authorities under the Treaty. The trawler owners are now claiming damages against the Department of Transport.

The four freedoms of the Community

The four freedoms of the Community, covered in Part Two of the Treaty under the heading 'Foundations of the Community' along with provisions on agriculture and transport, are the free movement of goods, free movement of persons, freedom to offer services and freedom of movement for capital.

Other areas of interest are in Part Three of the Treaty, headed 'Policy of the Community', which includes rules on competition, tax provisions, approximation of laws and social policy.

Free movement of goods

The basis of the free movement of goods is Article 30 of the Treaty which reads as follows:

> 'Quantitative restrictions on imports and all measures having equivalent effects shall, without prejudice to the following provisions, be prohibited between Member States.'

This has been elaborated in the 'Dassonville formula' (*Procureur du Roi* v. *Dassonville* [4.23]):

> 'All trading rules enacted by Member States which are capable of hindering, directly or indirectly, actually or potentially, intra-

Community trade are to be considered as measures having an effect equivalent to quantitative restrictions.'

Article 36 provides derogations or exemptions, when prohibition or restrictions may be justified:

> '. . . on grounds of public morality, public policy or public security; the protection of health and life of humans, animals or plants; the protection of national treasures possessing artistic, historic or archaeological value; or the protection of industrial and commercial property. Such prohibitions or restrictions shall not, however, constitute a means of arbitrary discrimination or disguised restriction on trade between Member States.'

These exemptions have been interpreted narrowly by the Court of Justice. Generally, goods which can be sold in one Member State cannot be prohibited in another Member State even when national law appears to prohibit the sale of the goods. A good illustration of this principle is the well known *Cassis de Dijon* case [4.24] concerning a French blackcurrant (cassis) liqueur which did not contain the level of alcohol specified in German law for a liqueur. The Germans alleged, to no avail, that the measure was enacted in the interests of public health since it was designed to prevent the increased consumption which would result if the alcohol content was reduced! (This reason stands comparison with a Japanese attempt under GATT to exclude a certain brand of mints on the grounds that their particular shade of green was harmful to Japanese eyes.)

The principle that national rules against products should only be permitted for a very limited range of reasons is incorporated in the 'new approach' to products standards. The 'new approach directives' now generally restrict themselves to specifying the essential requirements, which are primarily concerned with safety and health, and the environment.

Free movement of persons

Free movement of persons is not as yet general within the Community. Individuals and their families are free to work and look for work throughout the Community (this does not, however, apply to Spanish or Portuguese workers until 1 January 1993). The Commission has for some time been working on proposals to extend the freedom of movement to students, the retired and others. The main difficulty in reaching agreement in the council is that the Member States which offer high levels of social benefits, particularly those in the north of the Community, are concerned not to make it possible for others to exploit these benefits.

There is also concern about well-paid workers from the north being undercut by imported cheap labour from low wage Member States such as Portugal or Greece. In *Rush Portuguesa Limitada* v. *Office National d'Immigration* [4.25] the court confirmed that a Portuguese construction contractor could carry out work on French soil using Portuguese labour who could work under the same conditions as those imposed by the host state on its own nationals, which meant, in effect, without permits. Mobility of labour, geographically and between industries, is a key mechanism for allowing the internal market to operate efficiently, even though this may be unpalatable to some.

Permits should not be necessary, but Member States may legitimately require registration for those exercising their freedom of movement. The individual and his or her family should be entitled to the local conditions applying to pay, tax, working conditions, trade union rights and social security.

Member States may impose restrictions in the interests of public policy, public security or public health, which terms have been elaborated by Council Directive 64/221/EEC [4.26]. Applying the principle of proportionality, only factors affecting the fundamental interests of society are sufficient to justify exclusion. The court overruled a French decision to restrict the movements of a noted political agitator from Italy, but upheld that a Dutch citizen who intended to take up employment in connection with a college of Scientology in the UK could validly be excluded.

Certain listed diseases or disabilities may be used to justify exclusions under the same directive, or to justify the refusal to grant a first residence permit.

Freedoms of establishment, services and movement of capital

Freedom of establishment (Articles 52–58) includes the right to 'take up and pursue activities as self-employed persons and to set up and manage undertakings . . .' It does not apply under Article 55 of the Treaty to 'activities which in [a Member] State are connected, even occasionally, with the exercise of official authority'. The introduction of measures providing for mutual recognition of qualifications is provided for under Article 57. The Single European Act brought in the co-operation procedure, and therefore qualified majority voting in this context. The Architects' Directive had taken 17 years to agree so a change was clearly needed! Under Article 58 companies and firms registered in any Member State 'shall . . . be treated in the same way as natural persons who are nationals of Member States'.

Articles 59 to 66 relate to the provision of services and the removal of restrictions. Transport is treated as a special case, and has a section of its own (Articles 74 to 84). Some of the barriers in this protected sector are only gradually being tackled.

Articles 67 to 73 deal with capital movements. Member States undertake to abolish restrictions on capital movements amongst each other and any discrimination based on nationality or on the place of residence of the parties or on the place where such capital is invested. In case large movements of capital cause problems the Member States do retain, for the time being, the right after consultation to introduce protective measures.

Chapter 5

Competition Policy, Company Law and Small and Medium Firms

The community operates a highly developed competition policy which may influence the activities of construction companies and companies in the building products sector. This chapter outlines the scope of Community competition law and company law. One particularly interesting development is the introduction of the 'European Economic Interest Grouping', a type of entity created to assist cross-border joint ventures and co-operation agreements. Community initiatives designed to help small and medium enterprises, which may be helpful to firms operating in the construction industry, are also described.

Competition

Under the chapter heading 'Rules on Competition', treaty Article 85 prohibits certain types of restrictive agreements and practices, Article 86 prohibits undertakings or groups of undertakings abusing a dominant position in the market and Article 92 limits the aid that states may give industry.

In this area the Commission, specifically DGIV, becomes directly involved in approving mergers and other agreements between companies, and may take legal action against undertakings which have infringed Community legislation. The Commission has the power to request information, to arrive possibly unannounced to question and to see files (with the assistance of the relevant authority of the Member State), and to impose fines of up to 10 per cent of turnover.

Jurisdiction is claimed over companies not established in the Community provided the Community market is affected. Under Article 85 various categories of agreement which hinder competition are prohibited and if prohibited are automatically void. There is a let-out provision in 85(3) which in some circumstances exempts agreements which bring benefits, a 'fair share' of which are allowed to consumers.

Article 85 and 86 refer to agreements and practices which may affect trade between Member States, but even if such agreements are restricted to one Member State they may come within the scope of the articles if the effect could be to restrict cross-border trade. In *Cementhandelaren* v. *Commission* [5.1] the Dutch cement dealers association regulated the price of cement in Holland. This was held to be likely to limit economic interpenetration, i.e. the ease with which foreign suppliers could penetrate the national market, and was therefore an infringement of Article 85.

'Gentlemen's' agreements and concerted practice, for example where a group of companies act in unison over price increases, have been held to infringe the treaty.

Agreements between companies who supply the same product or service (horizontal agreements) are clearly within the scope of the Article 85, but it has also been applied to agreements between suppliers and distributors – so-called vertical agreements. Thus Tippex, the manufacturers of correction fluid, were fined ECU 400 000 for trying to prevent distributors, who had been granted exclusive rights of distribution in one geographical area, from trying to sell into other areas. Actions of this sort by the Commission have brought criticism that it is too concerned with trying to bring about freedom of circulation of goods within the single market at the expense of overall economic efficiency. Why should one worry about such practices if they help companies to operate more effectively? Some economists argue that, provided it is possible for firms to break into a market, companies should be largely free to organise their operations without hindrance. It is possible that the Commission, seeing the increasing globalisation of markets and strong competition particularly from the USA and Japan, may take a more relaxed attitude to certain types of anti-competitive behaviour. It may even encourage firms to form mergers or to co-operate, in research and development for example.

The Court has accepted that in some circumstances an agreement which protects a market is acceptable. For example, in the maize seeds case, *Nungesser* v. *Commission* [Court of Justice 1978] [5.2] an agreement which protected the rights of a plant breeder was permissible, because without it there would be no incentive for bodies to invest in developing new breeds of plant. Justification of this type is referred to as the 'rule of reason', a concept which would be welcome in English law.

The Commission may explicitly exempt an agreement from the effect of Article 85(1), though the procedure is said to be time consuming and tedious. An example of an exemption, referred to in chapter 4, is that relating to the agreement between Eurotunnel and the 10 contractors, all major, who form the Transmanche Link consortium, for the design, supply, construct, test and commission contract for the Channel Tunnel. The Commission decision also covered the *maître d'oeuvre* contract with W.S. Atkins and Partners and Société d'Etudes et Economiques. The legal assessment in the Commission decision concluded that 'the construction and *maître d'oeuvre* contract do not have as their object or effect to restrict competition in the common market . . . and do not impose or imply any restriction on the freedoms of the member firms . . . as regards other activities'. Incidentally, the *maître d'oeuvre* contract covered the checking of designs, supervision of construction with reference to quality, cost and progress and acting as independent project manager on behalf of the Intergovernmental Commission set up between the French and UK governments to supervise construction and operation of the tunnel.

The Commission may issue a 'comfort letter' which indicates that it intends to take no further action. Another way that certain types of agreements may be deemed to be exempt is under a series of regulations which grant 'block exemptions'. Sometimes, the block exemptions are limited to agreements between parties with a certain combined turnover. Block exemptions have been adopted in relation to agreements covering exclusive distribution, exclusive purchasing, specialisation, research and development, patent licensing and motor vehicle distribution.

Article 86 prohibits the abuse of a dominant position so far as it may affect trade between Member States. Construction companies rarely reach a position of dominance in the market because the nature of the work limits the opportunity to gain significant competitive advantage. The use of new methods or specialised equipment is limited because the work is largely done on site and not usually repeated, and the scope for economies of scale are limited. A dominant position might be attainable in some specialised construction sectors, such as tunnelling, but is much more likely in building products. Pilkingtons, for example, might have a dominant position in the UK market for glass and is building itself a strong position in the European market, mainly through research and development. Hilti, well known in the construction industry for their nail guns, was found by the Commission to be abusing its dominant position by tying the sale of nails to the sale of cartridges, on which it owned the patent (Commission Decision, *Official Journal* L65 11.3.88 p.19). Eurofix had thereby been excluded from supplying nails. The question of what constitutes the relevant market is one which can and has kept lawyers and economists richly employed. The market for bananas was held, after much deliberation, to be distinct from the market for soft fruit, whereas the tin can market was held not to be so distinct because it was relatively easy for manufacturers to switch to alternative packaging materials such as glass. The ease with which purchasers are able to switch to another product, and their loss of 'satisfaction' in doing so, are the keys to the definition of markets.

Dominance is not defined but a share of 35 per cent or more of the EC market might be regarded as indicating dominance.

It is possible for companies or individuals to complain to the Commission about the abuse of a dominant position – there is even a standard form, Form C, obtainable from the Commission for doing so. The Commission Decision, referred to in chapter 2, under which Boosey and Hawkes were obliged to supply parts to Brass Band Instruments Ltd or face a fine of ECU 1000 per day, is an example of the result of a successful complaint.

Dumping

A related problem of 'unfair competition' is that of dumping, which is the selling, normally of products, in the Community at an uneconomic price by an external producer. The rules are laid down by Council Regulation 2423/88 [5.3], which is based on Article 113 of the Treaty (common commercial policy). The Japanese and other Far Eastern producers have fallen foul of rules in numerous product areas. Many have accused the Community of using the anti-dumping rules as a form of protectionism. Under the Regulation the Commission and Council have the power to impose duties on imported goods. Electronic components such as 'chips', and goods, such as dot matrix printers, have been the target, but construction plant imports have also been affected. For example, the EC imposed duties on Japanese crawler-excavators between six and 35 tonnes for a five-year period. This ended in 1990 and was not renewed. Probably as a result of the duty, Japanese manufacturers have started building excavators in the Community and around 85 per cent of Japanese excavators now sold in Europe are said to be built in the Community. Komatsu has a plant at Birtley in the UK, and Hitachi has a joint venture with Fiat in Italy.

State aids

The touchy question of state aids to industry, often granted to industries which are in decline or regarded as strategic, is dealt with in

Article 92. In general, state aid is incompatible with the common market if it 'distorts or threatens to distort competition . . . in so far as it affects trade between Member States'. Certain aids are declared compatible with the treaty, for example if they are 'social' in character and 'granted to individual consumers . . . without discrimination . . .' or to make good the damage caused by 'natural disasters or exceptional circumstances'. Other aids *may* be compatible, for example to 'promote the economic development of areas where the standard of living is abnormally low or where there is serious underemployment'. The Commission, again in the form DGIV, is charged with monitoring this aspect of the internal market. Sir Leon Brittan, the Commissioner with responsibility for competition, has steered the Commission into taking a tougher line on state aids. Both the British and French governments have had problems with the Commission in this area: Britain in connection with 'sweeteners' in the sale of Austin Rover to British Aerospace, and the French in connection with subsidies to state owned industry. In France, state owned companies, such as Renault, contribute 30 per cent of the gross national product, and the state has a tradition of interventionism.

British offshore fabricators are among those who feel threatened by state aids, particularly in combination with the public procurement utilities directive (see chapter 6). Apart from concern about competition with yards with access to cheaper labour, yards in Spain, Portugal and Italy also receive state support. These are in areas where the standard of living is abnormally low and aid 'may be considered to be compatible with the common market' under Article 92(3).

One area unfortunately not within the scope of the Article 92 ban is state support for exports. British contractors and consultants complain that other countries provide more support, usually in the form of cheap export credit guarantees or cheap project loans. For example, a German consortium led by AEG and Siemens won the first stage of an underground railway project, which was financed by a $220 million loan from the German government, for Shanghai, China's second largest city.

Control of mergers and other 'concentrations'

Until 21 September 1990 there was an awkward mix of national regulation of mergers, and Commission involvement. But from then, certain categories of 'concentrations' have been primarily the responsibility of the Commission, again in the form of DGIV [5.4]. A concentration is when two or more previously independent undertakings merge, or where direct or indirect control is obtained by purchase of shares or assets, by contract or other means. A proposed concentration must be notified to the Commission if the proposed concentration will bring worldwide aggregate turnover of all the undertakings concerned to ECU 5 billion or more and aggregate turnover of at least two of the undertakings concerned to more than ECU 250 million within the Community. However, if more than two thirds of the gross Community turnover is within one Member State it is then the Member State, not the Commission, which has responsibility for examining the proposed concentration.

Company law

Most of the long series of directives and proposals on company law are concerned with standardising accounts and other information provided to the public. There is also a proposal on procedures for informing and consulting employees (the 'Vredeling' directive) which has long been blocked in the Council, and one on the structure and management of public limited companies employing more than 1000 people in the Community. The latter proposal, which would have required some form of employee representation, is shelved for the moment, but aspects of it may be introduced into future proposals.

The 'twelfth company directive' [5.5], which is due to be implemented by 1 January 1992 makes it possible to set up private limited companies with only one member. The UK currently requires at least two members. This measure is intended to help small companies.

Of possibly more interest to those concerned with undertaking work around the Community is the proposal for a European company statute [5.6]. Agreement in the Council on this proposal which is intended to help companies trading in several Member States is likely to take some time. Under the proposal the company, a Societas Europaea (or SE), from the Latin, would have a two-tier management structure providing for employee participation. Profits of any permanent establishment of the company in a Member State would be taxable in that Member State alone.

European Economic Interest Grouping

Another form of corporate body has already been introduced to facilitate cross-border activity, the European Economic Interest Grouping (EEIG) [5.7]. An EEIG is a body set up by bodies or individuals from at least two Member States and whose activities must be related to the activities of its members. The EEIG cannot employ more than 500 persons and profits must be distributed to its members. At least one firm of architects has become a member of an EEIG and claims to have derived considerable benefit. They are said to be suitable for low risk ventures involving businesses such as law or accountancy firms, high risk high value projects such as construction projects, and low risk 'horizontal groupings' whereby companies might set up common services, such as transport, research and development or marketing for the members. One disadvantage is that its members share joint and several liability for the liabilities of the EEIG.

Small and medium enterprises

The Commission has long been keen to help small and medium enterprises (SMEs), seeing them as a means of generating jobs and as important in developing the more depressed regions of the Community. SMEs are defined as having less than 500 employees, net turnover less than ECU 38 million and not more than one third of the ownership ascribed to a parent or financial organisation. They represent 95 per cent of all Community firms, but are rather less significant in volume terms than in the USA and Japan, both of which have active programmes to promote small firms. Many of the measures targeted at SMEs could be used by construction firms, though some of the measures are designed to promote new technology and may be most relevant to manufacturing industry.

Community competition policy has already provided some protection for small companies from anti-competitive behaviour and government support which distorts competition, but more active support was felt to be necessary.

In June 1986 the Commission created a 'task force' in DGIII, (the internal market directorate générale) with the specific remit to provide help to the SMEs. This has since been incorporated in a new directorate générale, DGXXIII, which has responsibility for enterprise policy, distributive trades, tourism and co-operatives.

Information and advice

The task force has set up a network of 187 Euro Info Centres (known in Brussels by the French term *Euroguichet*) distributed around the regions of the Member States. Their main task is to provide information and advice to small and medium-sized firms. The centres operate from within existing organisations such as enterprise agencies, chambers of commerce or professional offices, offer newsletters and local events aimed at raising awareness of opportunities, and advice, consultancy and information. They have access to expertise in Brussels and can obtain answers to questions about other Member States as well as about Community programmes. They are linked to electronic Community databases, including TED (*Tenders Electronic Daily*) which contains information about public and other contracts, Celex (the Community law database), press releases and Community funded research and development.

Another source of information and advice is the network of 24 Business and Innovation Centres which are aimed at regions hit by the decline in industries such as coal and steel. These centres can also provide start-up premises and training.

Making links and finding partners

To help companies find partners for joint ventures, mergers, co-operation, joint research, or for investment, an electronic business match-making service called BC-Net has been created. Again the Euro Info Centres can provide access. In November 1990 the database held 25 000 profiles of companies. Another initiative is the 'Partenariat' scheme which brings companies, identified via BC-Net, from depressed regions and potential investors into contact. The first event under this initiative took place in Dublin and was said to have had a 60 per cent success rate. The second was in Torremolinos in Spain and involved more than 220 local businesses and about 150 others.

Other help for small and medium firms

The Commission has set up schemes for providing capital, particularly 'seed capital' to help small companies get new ideas and products launched. Loans at preferential rates are also available through the New Community Instrument. This is administered through national banks via the European Investment Bank. Another strategy, is aimed at reducing bureaucratic burdens on small companies; this has resulted in a simplified VAT scheme for them, and lighter requirements for annual company reports and audits. However, some Member States, including the UK, have decided that the annual audit should remain a requirement. Since 1986 the task force has assessed all Commission proposals for their impact on SMEs and has had a number of changes made. Small and medium enterprises are given special treatment under the Community research and development funding schemes such as SPRINT, ESPRIT and BRITE and also under the European Regional Development Fund and by the European Investment Bank (see chapter 12). Firms with less than 25 employees are given preference under schemes for education and training funded by the European Social Fund.

A more recent scheme, launched in July 1990, was for the 'preparation of regional industry for the single market', inevitably referred to by an acronym, in this case PRISMA. A sum of ECU 100 million was to be made available from the European Regional Development Fund under the scheme. Member States were invited to submit proposals for improving calibration, testing and certification facilities in the depressed areas, and for providing help for small companies in finding out about, and in tendering for, public contracts. Help is to include information about quality systems, legal and procedural obligations in other Member States, about research and development results relating to products, new market opportunities, and the organising of exhibitions, seminars or conferences related to public procurement opportunities. Financial help was also to be provided, for example in relation to registration and admission to lists of qualified tenderers, translation and other costs involved in tendering for work outside a firm's local area.

The wide array of support for the small and medium companies and much other information about the Community is most easily accessed through the local Euro Info Centre.

Chapter 6

Public Procurement

OVERVIEW

Opening up public procurement to greater competition from non-national firms is regarded as a key component of the completion of the internal market. The White Paper on the completion of the market contained two and a half pages discussing its importance and describing the measures needed to achieve opening. During the main development period of the proposals a separate directorate dealing with the development of new proposals and the enforcement of existing directives was created in the Directorate Générale concerned with the Internal market, DGIII. This was under the leadership of a British barrister, Robert Coleman, who had a record of achievement in other areas. The Directorate has now been subsumed into DGIII/B which has responsibility for the 'horizontal' aspects of the internal market, i.e. the measures which may affect all industrial sectors. Apart from public procurement this directorate deals with standardisation and the free movement of goods (Article 30 *et seq*.). The section dealing with new initiatives in the area of public procurement is led by David White.

Business perception of the importance of barriers

Despite the Commission's emphasis on public procurement a survey on business perceptions of market barriers in the 12 Member States ranked the following ahead of government procurement practices:

- administrative barriers;
- differences in national standards and regulations;
- regulations of freight;
- restrictions in capital markets;
- community law.

Only differences in VAT were rated lower. In three countries government procurement practices were perceived as of higher ranking: in Italy it was regarded as the second most important barrier, in Portugal it was ranked third and in the UK fourth. Since the public sector commissions a substantial proportion of new building and the majority of civil engineering work, public procurement must be a key component of the single market for the construction industry. Public procurement practices, which are often very nationalistic or even regionally biased, must be tackled if there is to be anything approaching a single market, though not without attention to standards, regulations and practices, which can also create problems.

Benefits of opening up public procurement

The study on the 'cost of non-Europe'

What are the potential gains from opening up public procurement? The 'Cost of non-Europe' study referred to in chapter 1 concluded that, even without the effect of medium term economic measures to accompany the completion of the internal market, the benefits from opening up public procurement would amount to 0.5 per cent of Community GDP and lead to 350 000 additional jobs. This was out of a total growth attributable to a fully barrier-free internal market of 4.5 per cent of GDP and 1.8 million extra jobs. The benefits would be even greater if the completion of the internal market was accompanied by the medium term economic measures.

Savings related to public procurement

Parts A and B of volume 5 of the 'cost of non-Europe', comprising 828 pages between them deal with public-sector procurement. It provides full details of the data gathered and the techniques of economic analysis used. 165 pages are devoted to analysis of the public sector and purchasing procedures in the five countries on which the study was based – Belgium, France, Germany, Italy, and the UK. Much of the report deals with the potential impact of opening up procurement in the water, energy, transport and telecommunications sectors. At that time these sectors were excluded from the public procurement directives, and large potential gains and savings were apparent, particularly from the restructuring of the suppliers and mechanical and electrical contractors to the industries concerned. Substantial economies of scale could result in the production of power generation equipment, for example. Current diseconomies demonstrated were in many cases very striking.

Evidence of nationalistic purchasing practices

General import penetration for the five countries studied ranged from 43 per cent for Belgium to 19 per cent for Italy. In marked contrast, public sector import penetration of foreign goods ranged from 21 per cent in Belgium to 1 per cent in Italy. Public purchasing from foreign suppliers and contractors was, not surprisingly, even lower, ranging from 3.8 per cent in West Germany to 0.3 per cent in Italy.

The volume of public procurement

The study estimated that total public sector purchasing in the EC was about 15 per cent of GDP (total public sector expenditure is much larger and includes non-purchased items such as wages, financial charges, grants and subsidies). Of public purchasing about 15 to 25 per cent was non-competitive or non-tradable services or products such as electricity, rents and postal services, while another 20 to 30 per cent was for small items or other purchases not subject to formal contract procedures and not likely to be bought abroad.

However valid the study was for other sectors, the section on building and civil engineering was questionable. These sectors are said to represent about 30 per cent of public purchasing but only about one and a half pages are devoted to the sectors and half of that is spent on an account of the tendering and costing of one of the Commission buildings. The section concludes that there is only limited scope for savings, primarily in border areas, in specialist activities such as tunnelling, airports, and innovative building methods, and finally when construction prices in one Member State are high because the industry is over-stretched and while that in other Member States has spare capacity. The most contentious aspect of the study in relation to building and civil engineering is the estimate of potential 'static' savings, i.e. the savings without restructuring or the 'competition effect' (the driving down of prices charged by domestic producers because of increased foreign competition). The 'Atkins hypothesis' for this is a bland assumption that 10 per cent savings would be possible. The only apparent evidence for this hypothesis is that the value of the contract for the Commission, the only contract studied, was rather less than the 1981 prior estimate made by the Belgian Ministry of Works.

However, and with justification, no potential savings were shown resulting from the competition effect, or from the restructuring effect (the changes in the size or arrangements for production to take advantages of economies of scale, research and development advances, etc.). It is certainly true that the industry is unlikely to reduce costs much by firms becoming larger. The report comments that there might be restructuring benefits in a number of specialist areas such as nuclear installations, ports and harbours, but no savings are quantified or included in the estimate of overall benefit.

An across-the-board saving of 10 per cent in building and civil engineering is not credible, given the already highly competitive nature of the industry, the low levels of profit on turnover which are normal, and the fact that so many of the contracts carried out are too small to interest foreign contractors. Local labour will be used in all but the most special circumstances precluding any labour cost advantage, and the scope for changing the design of works to introduce more economic methods of production is small. The situation when the construction industry in one Member State is much busier than in others is rather unusual, and probably short-term. There was a brief period in 1988–89 when the UK industry in the south-east was over-stretched – but even then there was spare capacity in other parts of the country. Little labour was imported to the south-east, mainly because of the expense and scarcity of housing near the work. Some savings are possible, but 10 per cent is too high.

Savings in procurement of business, engineering and architectural services

Another area where a potential 10 per cent saving is assumed is in business services, which are shown as accounting for 4.63 per cent of total public purchasing. 'Business services' includes engineering, design and architectural services along with others such as management consultancy, computer services, insurance and finance brokers, estate agents, legal services, accountants and advertising. The 10 per cent estimate of potential savings is made despite the acknowledgement that language, knowledge of local law, customs and circumstances are important barriers to cross-border trade in such services. In addition, in some countries the price charged for certain services is fixed by law and no competition on price is possible. This applies to architectural services in Germany, for example.

Legal basis for action in the public procurement area

There are no articles in the Treaty specifically concerned with public procurement, but several are linked and are used as the legal basis for directives in this area. Governments must follow the requirements of the Treaty in awarding contracts even if below the thresholds. For example, Article 7 provides that 'within the scope of this Treaty, and without prejudice to any special provisions contained therein, any discrimination on grounds of nationality shall be prohibited'. The Treaty chapter entitled 'Approximation of laws' is relevant because in most Member States there are laws regulating procurement procedures used by public sector bodies. Article 100a, the article dealing with the creation of the internal market and providing for qualified majority voting in the Council of Ministers, is clearly relevant. In the matter of freedom of movement of goods and the specification of products for works, Article 30, which prohibits measures having an effect equivalent to quantitative restrictions, also has relevance. Since nationalistic purchasing by government bodies could harm the ability to establish and to provide services, Articles 52 and 59 also have a bearing on the powers of the Community in this area.

Some cases involving public procurement

Portuguese labour in France

In *Rush Portuguesa Limitada* v. *Office National d'Immigration* [6.1] a Portuguese construction firm won a sub-contract from a French company to carry out work on a railway line in the west of France. They brought in labour from Portugal to carry out the work, but in breach of the French *Code du Travail* (labour code) relating to foreign workers, and were told that a special charge would be payable. Legal action was taken and the case was referred to the European court. It was held that the Portuguese company was entitled to bring its own labour and that the French government could not impose conditions requiring that work permits be obtained. This judgment followed the principles of freedom to provide services under articles 59 and 60 of the EEC Treaty. The case was complicated because there were transitional rules applying to Portugal which did not expire until 1 January 1993. However, interpreting these rules in the light of their purpose, which was to prevent a sudden and substantial movement of workers, the derogations from the full application of the Treaty did not apply to the temporary movement of limited numbers of workers to carry out specific work for their employer. Nevertheless the judgment recognised that Community law did not prevent Member States from applying their legislation or collective labour agreements to any person working on its territory, even if temporarily.

The 'Dundalk' case

An example of the application of the Treaty to public procurement came in *EC Commission* v. *Ireland* (*Dundalk* case) [6.2]. The dispute concerned a contract let by the Dundalk Urban District Council for the construction of a water main. The tender enquiry documents contained the following clause:

> 'Asbestos cement pressure pipes shall be certified as complying with the Irish Standard Specification 188:1975 in accordance with the Irish Standard Mark Licensing Scheme of the Institute for Industrial Research and Standards. All asbestos cement water-mains are to have a bituminous coating internally and externally. Such coatings shall be applied at the factory by dipping.'

One of the Irish contractors tendering for the scheme submitted a tender which included for pipes manufactured in Spain to the international standard ISO 160–1980, but not certified as complying with the Irish standard. In fact only one firm manufactured pipes which were certified as required. The tender proposing Spanish pipes was rejected because the pipes did not comply with the specification clause. The contractor and Spanish pipe manufacturer complained to the Commission, who took the Irish government to the European Court following the procedure laid down in Article 169 of the Treaty. The court held that the Irish government had failed to meet its obligations under Article 30 of the Treaty. The inclusion of the words 'or equivalent' in the specification together with proper technical consideration of alternatives would have been sufficient to meet the obligation.

In other circumstances the Irish government would have had to pay the other parties' costs, but the Commission had failed in another of its contentions. This was that the Irish government, which had chosen to publish a tender notice in the *Official Journal* of the European Communities had thereby chosen to be bound by the requirements of the public works directive 71/305/EEC, even though contracts let by services concerned with water distribution were excluded by Article 3(5) of that directive (see p 87). Not surprisingly the court rejected this proposition and accordingly made no order for the reimbursement of the Commission's costs. If the proposed directive on remedies in the utilities sectors(1990) [6.3] (see p 109) had been law at the time, the wronged contractor could have asked for intervention in the award procedure and could possibly have claimed damages from the contracting authority.

Differences in legal approach amongst Member States

In the UK there is no legislation relating specifically to the contractual procedures used by public bodies. Internal control and audit are used to ensure that value for money is being obtained; corruption is dealt with under criminal law, but within these constraints there is almost complete freedom to use the form of tendering procedure and contractual arrangement thought to be most suitable for the procurement of works, supplies and services. There is no distinction between private and public bodies in the law of contract.

There have been various reports in the UK examining alternative contractual arrangements and tendering procedures for public works, the last being the Wood report, or more properly *The Public Client and the Construction Industries* [6.4]. This report gave recommendations (note, not instructions) on a great variety of methods of contract letting and management. One procedure strongly advised against was the open tendering procedure widely used in some Member States and permitted in all the public procurement directives and proposals. This recommendation was derived from research which showed the procedure to have a poor record on both cost and completion on time. It also has other disadvantages such as the high costs of abortive tendering which the industry has to pass on to the client and is slow and costly to administer if many firms respond.

Because the procedures are not regulated by law, there has been very little remedy available if a contractor feels himself to be unfairly excluded or treated. Most invitations even contain the warning that there is no obligation to accept the lowest or any tender. The comments about the lack of legislation refer to law originating in the UK. There have, of course, been the EC directives, though these have been implemented only by means of instructions issued to the relevant bodies by the Treasury.

The situation is very different in most other Member States, where it is normal for procurement procedures to be regulated by law, and often in great detail. The directives on public procurement build on this tradition, and practice in the UK will soon be harmonised to the extent that it will be possible to claim damages from a tendering authority which infringes Community rules and thereby adversely affects a contractor's chance of being awarded the contract. A brief overview of procurement procedures in some of the other large Member States will serve as a useful context for the public procurement directives.

Belgium

In Belgium the law covers works, supplies and services. In most cases competition, and a fixed price, are required. The law applies to all ministries and regional bodies down to provincial and communal level. It also applies to nationalised companies which are totally subordinate to a ministry, such as RTT (the telephone company), and SNCB (the national railway company) which act like private companies even though wholly owned by the state.

Five procedures are recognised: the *l'adjudication publique* and *l'adjudication restreinte,* both of which are based on price alone, the former with open tendering, and the latter restricted tendering; *l'appel d'offres, général* and *restreint,* which are similar to *adjudication* but take quality and performance into account as well as price and finally *le gré à gré* or negotiated procedure. The authorities have discretion amongst the first four, but are only allowed to use the negotiated procedure in defined circumstances, for example in case of urgency or where there is only one company able to do the work.

For construction works there is a register of contractors qualified to carry out work in particular categories, up to certain maximum contract values, from which tenderers must be selected (with special rules applying for contractors from other Member States). The subject of national rules on the pre-qualification of contractors will be referred to later.

France

In France there is a very clear set of regulations governing the procedures associated with public contracts, the *code des marchés publics.* In the Latin system of law, where there is no developing body of case law, codes are used instead to set down the law, and must be periodically revised to take account of developments. The code is therefore law and not mere guidance. It is used together with more detailed rules relating to specific aspects of the procedure – various *cahiers des clauses,* but the conditions of contract themselves, the *cahiers des charges*, are not so closely regulated. The regulations apply to all central government purchasers, to local authorities and their public bodies, and to public bodies other than those with a commercial or industrial character such as SNCF (the national railway company), Electricité de France, Aeroports de France and so on. In practice these companies tend to follow the procedures.

The procedures are similar to those in Belgium. The *l'appel d'offre*, where there is competition and tenders are evaluated on price and quality is the most commonly used procedure. Again negotiated contracts are only allowed in certain circumstances, very much as in Belgium. There is a commission with responsibility for supervising and updating public purchasing, the *Commission Centrale des Marchés*, which also issues statistics about purchasing in France.

The code is said to be comprehensively drafted, but to result in bureaucratic checks and delays. Tendering authorities manage to find ways around the system, for example by specifying the requirements in such a way that only the favoured firm fits. The system does not require reasons for rejection to be given, an aspect of what is referred to in Brussels as 'transparency', and which the recent family of EC public procurement directives is partly designed to remedy. The British suspect the French of chauvinism, and research seems to confirm that this is applied to purchasing, where possible. There is no official pre-qualification system, but the need to provide the legally required 10-year insurance for works has had the effect of requiring contractors to be certified by OPQCB (*Office Publique pour la Construction des Bâtiments*), since the insurance companies offering the *décennale* require certification.

The code also deals with engineering services but is not so prescriptive as in relation to works. The procedure generally used is for contracts to be let by negotiation but with a prior call for expressions of interest. The code even deals with design contests, sometimes used for prestigious schemes such as the Centre Pompidou. Notwithstanding earlier comments about chauvinism, foreign architects seem to have been very successful in winning such competitions.

Germany

In Germany procurement procedures are not governed by law but are set out in federal, state and community budget regulations. The procedures to be used are described partly by reference to Verdingungsordnung für Leistungen (VOL) and Verdingungsordnung für

Bauleistungen (VOB). The latter covers construction work. Invitations to tender are published in various official journals. Open competition, restricted tendering and negotiated tendering are all provided for. Unlike the French system there is no procedure in which price alone is considered. As in the systems already referred to, negotiated tendering is only used in carefully defined circumstances. For construction, VOB covers the technical rules for the execution of the work. A characteristic of German procurement is the proliferation of small decentralised purchasing entities, with individual contracts therefore tending to be small and to come below the thresholds for coverage by EC public procurement rules. There is a tendency to select national firms for contracts, partly because of lack of confidence that foreign firms can meet the high standards required and also because of political considerations. There are also provisions within procurement procedures for favouring certain disadvantaged groups. Areas affected by the division of Germany had special status under the Treaty, and preference was given to regional firms (regional preference schemes are discussed at the end of this chapter).

Italy

Italy is characterised by a multiplicity of regulations stemming from a wide variety of central and regional authorities. It is said that compliance with all the regulations is impossible, because they are complex and often contradictory. Part of the Italian enthusiasm for European federalism may be that the Italians are sure that administration would become much more sensible and efficient. The unworkability of the regulations is matched by a creative genius in finding ways around the problems. Other characteristics are the strong political influence on the award of contracts, and in the award of planning approval. There are very large numbers of firms in the Italian construction industry, some of which, particularly in the south have links with crime. Construction is a convenient way of 'laundering' money and contracts may be 'bought'. Abnormally low tenders can thus be a problem for the authorities. Open tendering for contracts is usually required and very large numbers of tenders are common, sometimes 250 or more. The authorities have even taken to excluding tenders which fall more than a certain percentage below the mean as a way of reducing the numbers of tenders to be considered. It is also a way of excluding firms without risking personal contact.

The regulations govern more than just the tendering procedure, which they do in very great detail; they also specify the type of contract to be used. However, there are frequent cases where special local laws are passed giving exemption from regulations, and non-observance is in any case said to be widespread, with political or personal factors determining awards. One factor which used to influence awards was a requirement that all government and state entities place at least 30 per cent of the value of their purchasing with firms wholly or partially established in the Mezzogiorno (the southern part of Italy). This was one way in which government attempted to tackle the enormous economic and social disparity between the north and the south of the country. However, in Case C 21/88 *Du Pont de Nemours Italiana SpA* v. *Unita Sanitaria Locale No 2 di Carrara* [6.5] it was held that this regional preference policy was an obstacle to normal trade in the community and was contrary to Article 30 of the Treaty. Article 92, which sanctions amongst other state aids 'aid to promote the economic development of areas where the standard of living is abnormally low or where there is serious underemployment' and 'aid to facilitate the development of certain economic activities or of certain economic areas, where such aid does not adversely affect trading conditions to an extent contrary to the common interest' was of no avail.

Another feature of Italian procurement is the widespread use of concession contracts. These sometimes cover the granting of special rights to build some work and charge for its use, but it also refers to the delegation to a private company of a responsibility placed on an authority by government to provide a facility or service for the public. In some cases a concession may be more like what is known in the UK as a management contract with the works being handed over on completion. In other cases the concession extends to the management of the completed work with the right to extract revenue from it. The latter arrangement is similar to the build, operate and eventual transfer (BOT) form of arrangement familiar in the UK, for example as used for the Channel Tunnel and the Manchester Metrolink light rail project. The concession form of contract may be used for a range of works such as public housing, harbour works, schools, construction of state highways and the reconstruction and development of earthquake damaged regions. Tendering rules are meant to apply to construction-only *concessioni* and concession contracts over ECU 1 million, but are alleged to be frequently ignored.

Development of the public procurement directives prior to 1985

Public works contracts

Community legislation in the field of public procurement came into effect before the UK joined the Community with council directive 71/304/EEC [6.6]. This required Member States to abolish restrictions 'affecting the right to enter into, perform or participate in the performance of public works contracts on behalf of the state, or regional or local authorities or legal persons governed by public law'. It was published with directive 71/305/EEC [6.7] 'concerning the co-ordination of procedures for the award of public works contracts'. 71/305/EEC set out the basic framework of provisions which has been retained in all subsequent public procurement directives. The following describes the basic framework, though most of the provisions have been amended by 89/440/EEC as discussed later.

Definitions and exclusions

Basic definitions and exclusions from the application of the directive.

Technical specifications

Rules on the use of technical specifications; such specifications, which could make reference to national standards, to be issued with the contractual documents relating to each contract. This was to avoid foreign firms having problems in obtaining specification details. Trade names were banned except in specified circumstances.

Tendering procedure to be used

Only open or restricted tendering could be used, with certain specified exceptions, for example, when only one contractor was suitable for artistic or technical reasons, or in cases of extreme urgency brought about by unforeseen circumstances.

Advertisement and minimum periods

Rules requiring the publication of tender notices in the *Official Journal* of the European Communities, with minimum periods between the sending of the notice and either the receipt of tenders, or in the case of restricted tendering, a minimum period between the notice and receipt of requests to take part in the tendering procedure together with a minimum period to be allowed for tendering.

Rules on exclusion of firms

Restrictions on reasons which could be used for excluding firms from taking part in the tendering process.

Criteria for award

Rules on the criteria for the selection of a firm to be awarded the contract; this could be based on price alone, or to the 'most economically advantageous tender' when other criteria could be used including period for completion, running costs, profitability, and technical merit, for example.

Thresholds

Most of the provisions of the directive did not apply to contracts under 1 million units of account (equivalent to what is now termed ECU); there was a rule to prevent contracts being split with the intention of avoiding the application of the directive. The principle of using financial 'thresholds' has been retained in subsequent public procurement directives, despite arguments against it.

The excluded sectors

The directive did not apply to the public works contracts awarded by the 'production, distribution, transmission or transportation services for water and energy' or awarded by 'bodies . . . which administer transport services'. The reason for these exclusions was that entities in these sectors were not consistently in the public sector and it was felt unfair to have differing degrees of market opening among the Member States.

Entities providing telecommunications services were not, however, specifically excluded from 71/305/EEC. Even though telecommunications entities were not excluded, the water, energy, transport and telecommunications (WETT) sectors were later to become known collectively as the 'excluded sectors' until they ceased to be excluded and were instead referred to as the 'utilities' sectors (see p 100).

Public works concession contracts

The directive did not deal with concession contracts. Instead the Council made a declaration [6.8] requiring concession contracts over

ECU 1 million (then referred to as units of account) to be subject to advertising rules, with concessionaires being required to sub-contract at least 30 per cent of the value of the concession contract to third parties.

Supplies

Directive 77/62/EEC [6.9] was introduced to apply similar principles to public supply contracts. In this directive, however, open tendering was the norm. The threshold for application was ECU 200,000, the value being lower than that used for public works because it is easier to supply goods across borders than to carry out work abroad.

The GATT Agreement on Government Procurement

As discussed in chapter 1 the General Agreement on Tariffs and Trade (GATT) is agreed by over 100 nations, but, separate from this, is a sub-group whose members have agreed to the mutual opening up of government procurement of supplies. The agreement was negotiated within the context of the 1979 Tokyo round of negotiations between the then EC Member States, the USA, Canada, Japan, the EFTA countries, Singapore and Hong Kong. Since then Israel has joined and, of course, EC membership has expanded.

Directive 80/767/EEC [6.10] modified the supplies directive 77/62/EEC to take account of the GATT Agreement on government procurement. This explains why the *Official Journal* tender notices contain the occasional notices from government authorities well outside Europe. However the GATT government procurement agreement only applies to central government purchasing, and a different threshold is used, currently 130 000 'Special Drawing Rights' i.e. about ECU 134 000. The agreement only applies to goods, while the supplies directive also covers rental and leasing contracts.

Developments since the White Paper on the completion of the internal market

The White Paper proposals

The 1985 Commission White Paper on the Completion of the Internal Market listed seven measures in the field of public procurement, two of which have since been combined into the one 'utilities' directive.

Apart from the utilities which had been excluded from previous public procurement directives, the White Paper also referred to improvements in the directives applying to supplies and public works and to opening up the public procurement of services. The services associated with construction, such as architecture and engineering consultancy, were given special mention.

The White Paper commented that statistics indicated a minimal application of the directives. Less than 25 per cent of public expenditure in the areas covered by the directives was the subject of publication in the *Official Journal* and thus, even theoretically, open to Community-wide competition. There was a commitment to examine loopholes in the directives, such as the possibility of splitting contracts so as to fall below the threshold for application, to improve 'transparency of procedures' and also to improve the electronic database publishing system used for communicating tender notices (TED – *Tenders Electronic Daily*). Another commitment was to provide 'more visible action by the Commission in policing compliance with existing law' which would 'increase the credibility of the Community's efforts to break down the psychological barriers to crossing frontiers'.

In practice there had been widespread disregard for the requirements of the directives, partly through ignorance of their existence. Business was sceptical about the real openness of the markets and aware of other obstacles to competition, such as different standards. Even where notices were published contracting authorities reported virtual absence of interest from foreign firms. Firms were in any case not reading the notices.

Implementation of the proposals

There has been excellent progress on the White Paper proposals. Important improvements have been made to the works directive and the supplies directive. A new directive covers the purchasing by entities, whether private or not, in the utilities sectors (formerly the 'excluded sectors') i.e. water, energy, transport and telecommunications. A proposal for a directive covering the purchasing of services has been sent by the Commission to the Council, and a further proposal covering services in the utilities sectors is in preparation. A new directive provides remedies to enterprises for breaches of the requirements by entities covered by the works and supplies directives. There is also a formal proposal for a remedies directive in the utilities sectors.

In addition, very significant improvements to the coverage and scope of the GATT agreement on government procurement are likely to follow the completion of the Uruguay round, if that happens, and, finally, a section in the Commission has been active in dealing with alleged infringements of the directives and has harried Member States into following the procedures more closely. Some cases, such as the Danish Storebælt Western Bridge case (see p 109), have caught the attention of the trade press and sometimes even of the newspapers.

THE AMENDED PUBLIC WORKS DIRECTIVE

Directive 89/440/EEC [6.11] amending the works directive was adopted in July 1989 to be implemented 12 months later. Since the directive was a modifying directive the original directive on public works, 71/305/EEC, is still in effect though little of the original text remains intact. A consolidated version has been prepared, but at the time of writing had only been circulated to the advisory committees on public procurement. To understand the directive some form of consolidation is necessary and one imagines that any organisation which needs to refer to the provisions will have prepared its own version. The UK Treasury, for example, has done so and is willing to supply copies, together with guidance, on request.

The proposal for the modifying directive was dated 12 January 1987 and followed a study of attitudes amongst the Member States to the obligations imposed by 71/305/EEC. The amended directive was to cover contracts where the contractor did more than just construct, for example, where he carried out design, provided financing or project management. It was also to apply certain provisions to concession contracts. The use of single tender action was to be further restricted. It was to be made easier for contractors to find out about work in the pipeline by requiring contracting authorities to publish advance notices of their construction intentions. The threshold, which had remained unchanged since the original directive, was to be revised and the directive was to build on the developments taking place in the field of European standardisation. Another innovation was to require contracting authorities to give reasons for rejection if asked by a contractor to do so.

Summary of main provisions

Application

The directive uses the NACE classification 50 covering building and civil engineering work. This excludes mechanical installation works except those concerned with the installation of fittings and fixtures in buildings. All public bodies are covered, whether at state, regional or local level. State funded 'autonomous' bodies are also covered. In *Gebroeders Beentjes BV* v. *Netherlands* [6.12], the body concerned was a Dutch local land consolidation committee without legal personality of its own, but its function, composition and the means available to it for carrying out its functions showed that it acted on behalf of the state, or a regional or local authority. The directive therefore applied.

The procedures also must be followed by an entity which is funded by more than 50 per cent by a body which is covered, but only if the work involved is civil engineering work or related to hospitals, facilities intended for sports, recreation and leisure, school and university buildings and buildings used for administrative purposes. This wording evolved after much discussion because at least one Member State wanted to be able to fund house building without the award of construction contracts becoming subject to the directive.

Concession contracts

Concession contracts are subject to certain advertising rules if they are over the ECU 5 million threshold. Also, the concessionaire may be required to place at least 30 per cent of the contract value to sub-contractors, or the contracting authority can ask the tenderers to say what percentage of the contract they intend to sub-contract. The European Parliament wanted the authority to be able to use the percentage to be sub-contracted in assessing tenders on the assumption that a large percentage to be sub-contracted meant more work for smaller, and possibly local firms.

There are rules which in effect require the third parties, or sub-contractors to be truly independent, so that the rule cannot be avoided by the creation of a special holding company or other entity. Concessionaires are also required to follow some of the advertising rules before placing contracts over the threshold value to third parties. Certain bodies, particularly one representing French concession contractors, argued that concession contracts should not be subject to the advertising rules if the idea of the project had been originated by a particular contractor. The argument was that the idea should be treated as though subject to intellectual property rights which the owner should be given the right to exploit. Clearly if this had been allowed, there would be considerable scope for abuse. As the position stands, other firms may respond to the notice and put forward their own proposals. However, the originator of the idea would still have a head start which might be difficult to make up.

Exclusions

As before, the directive does not apply to bodies in the transport, energy or water sectors. However, the definitions were clarified to make it clearer that bodies providing facilities for transport, such as highway or airport authorities, were within the scope of the directive. Other exclusions apply to projects where state security is involved, for example certain military establishments, or works coming under other international agreements, for example concerning NATO or a UN agency.

Threshold

The threshold, after much argument, was set at ECU 5 million. There is now a mechanism for converting it into equivalent national currencies with values remaining fixed for two years at a time. (Note: as stated above public bodies must comply with the Treaty provisions even if contracts fall below the thresholds.)

Works divided into lots

Where the work is divided into lots the sum of the estimated values of the lots is the relevant figure to be used for applying the threshold, except that lots under ECU 1 million can escape the provisions of the directive so long as the sum of the exempted lots does not exceed 20 per cent of the sum of all the lots. This provision means that a project done on a construction management basis, where many trade contracts are placed, would not escape, and that if there were many small contracts, enough of them to constitute at least 80 per cent of the total would have to be subject to the rules. Some very small contracts may have to be advertised in the *Official Journal*.

Management contracts and nominated sub-contractors

The management form of contract which gained popularity in the UK during the 1980s is not mentioned, but the award of such a contract would seem to be subject to the rules, because the contractor agrees to be responsible for the execution of the work even though he is in effect mainly providing a management service. The individual works contracts would not, however, come within the scope, because they are not contracts with the contracting authority, even though the contracting authority is in a position to decide which sub-contractor should be employed. Thus, using this type of contract, the contracting authority could in theory cause most of the value of the work to done by firms of its own choosing, possibly on nationalistic grounds. It is likely that a purposive interpretation of the directive could negate this apparent loophole, which also applies to nominated sub-contractors. For example, it could be argued that the contractor in such cases would be acting as the agent of the contracting authority. In any case, the Treaty articles, such as Article 7 which prohibits discrimination on the grounds of nationality, still apply.

Splitting

In any case, work may not be split with the intention of avoiding the application of the rules. What the difference is between lots and any other method of splitting the project up is not clear. If the authority provides supplies to the contractor for the work, their value must be added to the estimated value of the contract before checking against the threshold. Though a reasonable provision, this could in exceptional circumstances result in some small contracts involving the installation of high value goods supplied by the client coming within the scope of the directive.

The award procedures

There are three procedures which may be applied: negotiated, with one or more firms selected according to qualitative criteria; restricted tendering where a limited number of firms selected from those interested is invited to tender; and open tendering where all those interested may tender. Contracting authorities may choose freely between the open or restricted procedures, but can only use the negotiated procedure in defined circumstances.

The use of the negotiated procedure

The negotiated procedure is sub-divided into negotiated with, and without publication of a tender notice. The negotiated procedure with publication is allowed if an open or restricted procedure produced only unacceptable tenders. The meaning of 'unacceptable' is restricted and would not cover the situation where all the tenders were considered too high.

Works for the purpose of research or development can be let by the

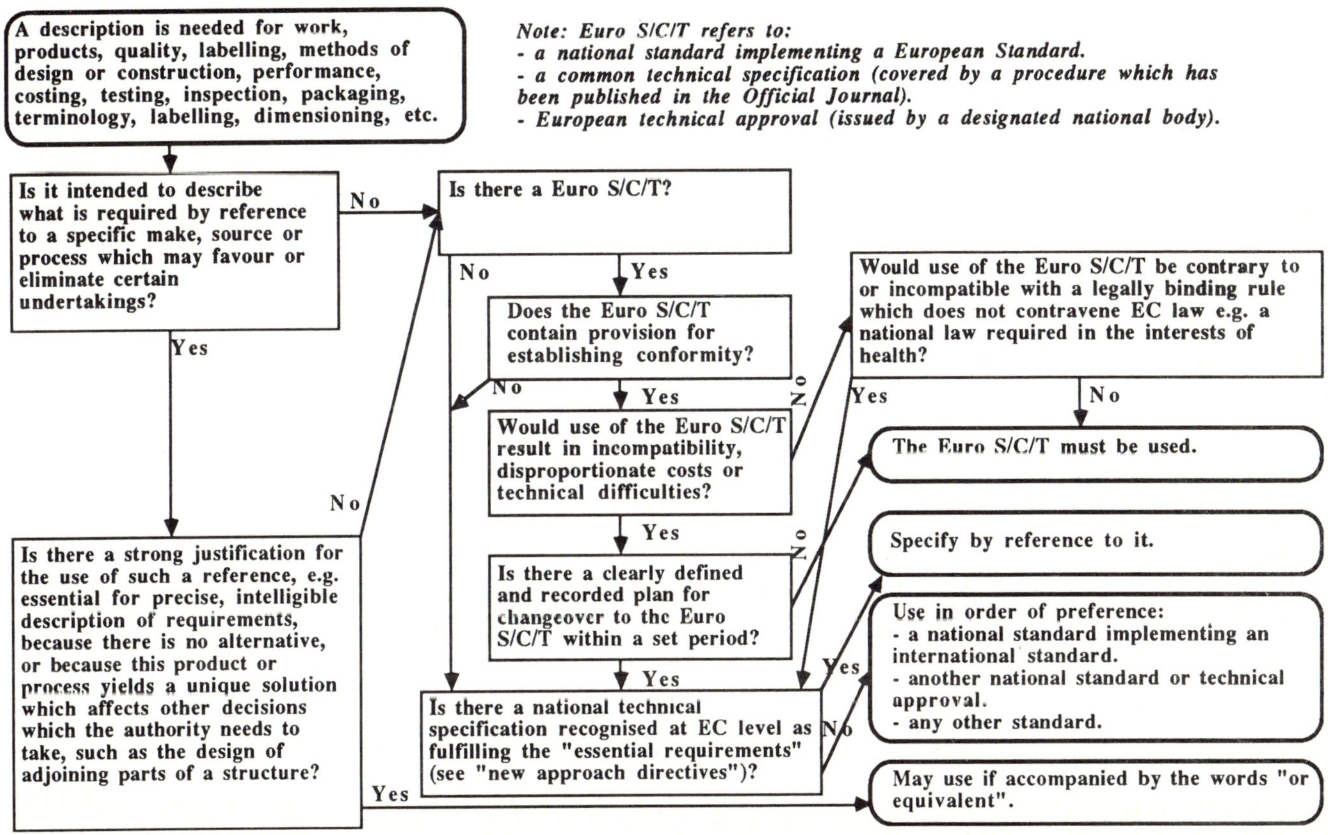

Figure 6.1 Public works directive rules on technical specifications for inclusion in contracting authorities, or certain contracts more than 50% financed by them, and of estimate value over ECU 5 million (or otherwise within the threshold).

negotiated procedure with prior publication. The purpose of the work itself must be research and development, for example, a trial section of road using a new form of construction. The construction of a section of conventional road which was to be used for vehicle research trials could not be let by this procedure. It can also be used 'when the nature of the works or the risks attaching thereto do not permit prior overall pricing', but only in exceptional cases. This might apply when an ancient building was to be carefully exposed and rebuilt as necessary.

The negotiated procedure without prior publication

The use of the negotiated procedure without prior publication of a notice is only allowed in the following cases:

(1) When there are no tenders or appropriate tenders following an open or restricted procedure, and provided a report is sent to the Commission, if requested. Statistics on the use of the procedures have to be sent to the Commission, following which further details might be requested.
(2) When for various reasons only one firm is suitable, or, for unforeseen circumstances there is extreme urgency (note that in all other languages the equivalent of 'unforeseeable' is used and this is likely to be the word a court would use).
(3) Where additional work is required and the original contractor is used, but only in certain circumstances.

Other provisions

Rejected candidates or tenderers have to be given reasons if they request them. This should generate care and scrupulousness in the authorities, particularly given the existence of the remedies directive.

There is special allowance for contracts relating to the design and construction of large or complex public housing schemes where early and close collaboration is needed with the contractor and others. Authorities may use a special but unspecified award procedure, with certain requirements applying to advertising in the *Official Journal*.

Technical specifications (see Figure 6.1)

The technical specifications have to be set out in the contract documents to make them as accessible as possible. The new Article 10 reads:

'. . .

2. Without prejudice to the legally binding national technical rules and insofar as these are compatible with Community law, such technical specifications shall be defined by the contracting authority by reference to national standards implementing European standards, or by reference to European technical approvals or by reference to common technical specifications.

3. A contracting authority may depart from paragraph 2 if:

(a) the standards, European technical approvals or common technical specifications do not include any provision for establishing conformity, or technical means do not exist for establishing satisfactorily the conformity of a product to these standards, European technical approvals or common technical specifications;

(b) use of these standards, European technical approvals or common technical specifications would oblige the contracting authority to acquire products or materials incompatible with equipment already in use or would entail disproportionate costs or disproportionate technical difficulties, but only as part of a clearly defined and recorded strategy with a view to changeover, within a given period, to European standards, European technical approvals or common technical specifications;

(c) the project concerned is of a genuinely innovative nature for which the use of existing European standards, European technical approvals or common technical specifications would not be appropriate.'

If the derogations provided for in paragraph 3 are used the notice in the *Official Journal* has to give reasons and the contracting authority may be required to give fuller information to the Commission.

However, the process of developing European standards for products will take many years. Nevertheless the Eurocodes (see chapter 9) currently being developed for the design of structures will have to be referred to in specifications once available and as appropriate. If the design is being carried out separately from the construction contract, this clause would not require that design to be according to the Eurocode.

Paragraph 10(5) covers the possibility that European standards, European technical approvals or common technical specifications do not exist. In which case the following apply:

- Specifications for products will have to refer to national technical specifications which satisfy the Construction Products Directive 89/106/EEC, due to be implemented in 1991, or other directives on technical harmonisation.

- If the specification needs to refer to design standards, or standards dealing with methods of calculation, to the way in which the work is to be done or how materials or products should be used then national technical specifications may be used. Thus for example British Standard Codes of practice, such as BS 5628 on the use of masonry, could be used. This could cause problems, since design and construction codes tend to make extensive reference to product standards. There could be considerable confusion if such a code is specified while a contractor chooses to use products meeting other national standards.

The last section of the article on technical specifications deals with proprietary products. Generally they should not be used except in cases where the use of the name is the only way of precisely and intelligibly describing what is needed. In that case the words 'or equivalent' must be used.

To overcome the effect of this virtual ban on proprietary names being specified in design there has been a tendency for UK public bodies to require the contractor to carry out more design work. There is, of course, no reason why the contractor should not specify proprietary products. (Some suppliers of innovative products which involved important design decisions had found themselves excluded from traditional work when the ban was implemented; Reinforced Earth Ltd was one such firm.)

Thus the British system, in which contractors are only invited to tender once the work is almost fully designed, specified and listed in the Bill of Quantities, has begun to change. It is moving towards the system used in France, where the tender drawings show only the concept design and the contractors are expected to carry out the detail design, for which they can use any proprietary products they wish, subject only to meeting the functional and, possibly, aesthetic requirements of the client. The impact of the directive is a convergence of national systems.

Variants

Where the criterion for the award of the contract is that of the most economically advantageous tender there are rules about the submission of alternative tenders. The contracting authority must state whether such tenders will be considered and state in the contract documents the minimum specification level to be achieved, together with the form of presentation required. Variants may not be rejected simply because they are based on technical specifications recognised or originating at the European level. This provision will be of little value until there are sufficient European Standards or national standards recognised as complying with the essential requirements, and this is likely to take many years. This article conspicuously does not refer to national technical specifications recognised as complying with the essential requirements as provided for under Article 4(3) of the Construction Products Directive (see chapter 9).

Advertising rules

Advertising in the *Official Journal* is required primarily to give contractors a chance to know of, and prepare for, work. It provides a source of information for work within, as well as outside, the contractor's own country. Having translations of tender notices may be helpful when contractors are assessing markets, but for works contracts, clearly local knowledge and particularly language will be unavoidable for tendering, let alone for carrying out the work. However, contractors may arrange translation if they have sufficient time or they may enter into some form of co-operation with a local firm. Also, the information will help firms to gauge the markets. Another function of the notices will be to enable the Commission to monitor various aspects of procurement; the amendments to the works directive will make the monitoring process much more effective.

Indicative notice, calls for interest or tenders, post award notice

Contracting authorities are now required to advertise as soon as the decision has been made to proceed with the planning of a contract – this is the indicative notice. Next, in the restricted procedure or the negotiated procedure with notice, a notice will have to be placed in the *Official Journal* calling for expressions of interest (the negotiated procedure without prior notice can only be used in a limited number of cases). In the open procedure a notice has to placed containing the invitation to tender. Once a contract has been placed, another notice has to provide details (in most cases). Apart from the details needed for assessment by contractors, the notices may also contain details about use of certain provisions intended for exceptional use, such as a decision not to use national standards implementing European standards, European technical approvals or common technical standards. This is to allow monitoring and possible follow-up by the Commission.

Contracting authorities wishing to place a works concession contract have to place a notice indicating this, and concessionaires themselves have to place notices referring to the award of contracts to third parties. Thus in a BOT (build, operate, transfer) project the contract for the BOT would have to be advertised, and then the successful contractor would have to advertise before placing major works sub-contracts.

Contracting authorities may advertise works contracts in the *Official Journal* even if the directive does not require them to do so. Since advertisements are free, this could be cost-effective, provided that sufficient numbers of contractors start using the *Official Journal*. The most convenient form of access is TED (Tenders Electronic Daily) which can be accessed by anyone with a suitably set-up personal computer and telephone line (see p 262).

Publication

Notices have to be sent to the Office of Official Publications in Luxembourg and, apart from those subject to the urgency procedure, are published within 12 days of dispatch. The indicative notice and the post-award notice get published in the nine official languages (translation care of the Office of Official Publications); other notices are published in the original language with a summary in the other eight. Publication is in the *Official Journal* (the S series), but TED would provide contractors with the maximum response time. Notices may not be published elsewhere before the date of dispatch.

Appendices to the directive contain mandatory pro-formas for the various types of notices, which may not be more than 650 words long. Contracting authorities are expected to be briefer than the legislators!

Minimum periods for response by contractors (*les delais* in French)

Minimum periods are set for certain stages in the procurement process to allow contractors not familiar with the procedures, or some distance away, to have a reasonable chance of responding. Periods are measured from the date of dispatch to Luxembourg. For the open procedure, 52 days (36 if the indicative notice was used) is the minimum permitted before the deadline for receipt of tenders. Since publication is allowed to be up to 12 days after dispatch the tender period could be cut down to 40 (or 24) days.

For the restricted procedure and negotiated procedure with prior notice, the periods are at least 37 days for expressions of interest to be received, and 40 days between the invitation to tender and the tender deadline. The invitations to tender must all be issued simultaneously. In all the procedures involving prior notices longer periods are to be allowed where site visits are essential for tendering.

When urgency makes the above time limits impractical there is a special procedure for publication of the notice within five days and shorter periods are allowed for expressions of interest (15 days) and tendering (10 days). (In the case of extreme urgency due to unforeseen (read unforeseeable) circumstances there is, of course, the negotiated procedure without prior notice referred to above.)

There are separate time limits for concession contracts and the contracts placed by the works concessionaire. Additional rules generally require quick dispatch of documents or additional information to interested contractors.

Qualitative criteria for selection or elimination of contractors

The directive allows only a limited number of reasons for excluding contractors in open tendering or eliminating them from those eligible to be invited to tender. These include bankruptcy, receivership or similar, professional misconduct, breach of a social security obligation or taxation requirements, or those 'guilty of a serious misrepresentation in supplying the information required under this chapter'.

Contractors can be assessed for financial and economic standing and for technical knowledge or ability, with guidance on what may be required as evidence. This may include financial references, professional and education qualifications, works carried out, tools and equipment, manpower and managerial staff and technical support. It is clear from the *Gebroeder Beentjes* case discussed below and the *Bellini* case (European Court of Justice, 1986) [6.13] that the assessment of financial or economic standing, or of technical knowledge or ability may include other considerations than those expressly listed.

Pre-qualification systems

There are also rules relating to enrolment on professional or trade registers and for overcoming national differences in this area. Some Member States, particularly Belgium, Greece, Italy, Portugal and Spain, operate official lists of approved contractors, usually subdivided according to categories of work or maximum size of contract for which approval is valid. The lists must be adapted so that they do not go beyond the rules on selection in the directive. Information which can be deduced about contractors because they are registered may not be questioned. For example, if only financially sound contractors were to be listed, then financial soundness would have to be presumed by contracting authorities in other Member States.

During the discussions leading up to adoption in the Council the Commission was asked to investigate the subject of official lists and make proposals if need be. A study has been made on pre-qualification systems in the Member States but as yet no decision has been taken about any follow-up action.

The problem raised by differing systems does not seem great. Contractors who had experience estimated that approval costs amounted to 1.25 per cent of tender costs, which themselves were only of the order of 0.5 per cent of the project value. Nevertheless, the administrative complications and delays introduced by having to deal with pre-qualification systems acts as yet another hurdle for contractors thinking about working outside their own national borders. Action by the Commission would probably be aimed at some degree of standardisation for all approval systems – even those operated internally by government departments, such as the Property Services Agency in the UK.

Procedures for reducing the numbers of tenderers

In the restricted procedure a minimum of five tenderers may be invited to tender, while an upper limit of 20 is suggested. No procedure is mentioned for selecting between qualified contractors, but any system chosen should not discriminate on grounds of nationality, or have the effect of disadvantaging foreign contractors. A criterion such as the distance from a firm's head office to the site in question would clearly be discriminatory. Presumably a genuinely random selection would be acceptable, but to demonstrate randomness the system would have over time to select foreign contractors in proportion to their relative numbers on lists. Another system might be to allocate places on the list of tenderers for foreign contractors in proportion to interest from such firms. Conversely, the system chosen should not favour foreign contractors at the expense of the national variety.

Contractors with a bad record of submitting claims

In 1989 the then Transport Secretary in the UK, Mr Peter Bottomley, threatened to exclude contractors from tender lists if they had a bad record for submitting claims. Such a practice would clearly be contrary to the rules on exclusion of participants. It would in any case be unreasonable to exclude contractors for legitimate application of the contract.

Politically based criteria for exclusion

At one time certain left wing local authorities started excluding contractors from their tender lists if they had carried out work in connection with nuclear bases, such as the cruise missile base at Greenham Common in Berkshire, and for other politically motivated reasons. This was dealt with at the time by a piece of national legislation, but 71/3 71/305 /EEC (as amended) would have prevented such criteria from being used on contracts to which it applied.

Criteria for award

Clearly objective criteria are needed to prevent authorities from rejecting bids on nationalistic grounds. Contracting authorities must decide in advance whether to select the cheapest tender, having rejected firms only on the basis of the qualitative criteria referred to above, or whether to select the 'most economically advantageous' tender. Examples of possible criteria for the latter basis of selection are given – price, period for completion, running costs, profitability and technical merit. However, if this basis for award is to be used the authorities must state either in the notice, or the contract documents, what the criteria are to be, and if possible, in descending order of importance. This could cause problems if unexpected factors result in new criteria being needed.

If there are pre-existing national rules preferring certain tenderers these can remain in operation, provided the rules are compatible with the Treaty, which is probably unlikely. The Commission must be informed about the rules.

Abnormally low tenders can only be rejected after the contractor has been given a chance to give a breakdown and explanation. If the lowest price is the basis of award, the Commission has to be told about the rejection of abnormally low tenders.

One interesting aspect is an exception to the rules on abnormally low tenders which can operate until the end of 1992. This allows the use of non-discriminatory procedures for rejection of tenders. This was incorporated to allow continued use of the Italian anti-Mafia procedures for the rejection of bids. In one procedure all bids below 5 per cent lower than the mean of all the bids are rejected.

Monitoring by the Commission

As part of the opening up of the internal market the Commission has been directed to develop its enforcement activities. To enable it to monitor the operation of the directive, Member States are required to send annual statistics about contracts awarded and procedures used to the Commission; Member States will have the burdensome task of organising a suitable system for collecting, collating and presenting these data.

When the derogation clause on the use of standards is operated the notice has to contain details and investigations may follow. Contracting authorities have to keep details about the use of the negotiated procedure without prior publication of a notice, and provide this information to the Commission on request (probably via national authorities). The other occasion when the contracting authority specifically has to inform the Commission is when an abnormally low tender has been rejected, but only when the criterion for the award of the contract was lowest price alone.

Member States also have to send information to the Commission, particularly about international agreements which result in works contracts being let in accordance with alternative procedures, and when there are pre-existing rules giving preference to certain classes of tenderer (e.g. from local firms), or when there are pre-existing schemes which affect the award of works contracts and whose objective is 'reduction in regional disparities and the promoting of job creation in regions whose development is lagging behind and in declining industrial regions'. These are permitted until 31 December 1992 'on condition that the provisions concerned are compatible with the Treaty, in particular with the principle of non-discrimination on grounds of nationality, freedom of establishment and freedom to provide services, and with the Community's international obligations.' (Article 29a – will be 31 in the consolidated directive.)

Social aspects covered in the works directive

Public works sometimes have a social purpose or are used in that way. They have been used in part to reflate the economy and provide employment, either nationally or to assist regions of high unemployment.

The Article 29a quoted above is clearly directed towards the social ends. The European Parliament made attempts to ensure that the directive took more account of these aspects but the final version does little more than pay lip service. The provisions themselves, which could have caused operational difficulties in some countries, have been diluted so that they are robbed of any power.

Article 20b (will be 20 in the consolidated directive) allows the contracting authority to ask the tenderers what share of the work is to be sub-contracted; an obscure subsidiary sentence states 'This indication shall be without prejudice to the question of the principal contractor's responsibility.' The parliament wanted firstly to enable contracting authorities to favour contractors who made considerable use of sub-contractors on the presumption that sub-contractors might well be local, and small or medium sized firms, and secondly, to ensure that the main contractor would be responsible for ensuring that sub-contractors comply with obligations relating to health and safety and local social conditions. Some members of the European Parliament representing constituencies in northern countries were particularly anxious that their national contractors and sub-contractors should not be undercut by outside firms with lower standards and lower rates of pay. Few traces of the requirements remain.

Article 22a (will be 23 in the consolidated directive) was inspired by similar concerns. It allows contracting authorities to provide information about employment protection and working conditions to tenderers, and allows Members States to make it obligatory to do so. If such information is provided tenderers must be requested to confirm that their tenders take due account of it. The original intention was both to help contractors gain access to the information, and to ensure that ignorance, wilful or otherwise, did not result in unfair competition with more responsible and better informed national firms. Denmark and Germany had particular concerns here, the former because wage agreements are given backing in the award of public contracts, and in Germany where there are regulations governing minimum wage rates.

In *Gebroeders Beentjes BV* v. *Netherlands* [6.14] the Court of Justice considered the rejection of a tenderer for a contract in which one of the conditions required at least 70 per cent of the work force to be from among the long-term unemployed and to be recruited through the regional employment office. Gebroeders Beentjes, who were rejected on the grounds that they lacked specific relevant experience and did not appear to be able to employ long-term unemployed, made a claim for damages in the national courts who referred certain issues to the Court of Justice. The above condition was not specifically examined but the court held that the list of factors which could be considered in assessing a firm's capacity contained in the public works directive was *not* exhaustive. It was held that any supplementary qualifications should be advertised in the notices which the directive requires (this requirement is not expressly stated in the directive). Any special pre-qualification requirements must not have the effect of discriminating against tenderers from other Member States.

It was held that any pre-qualification requirement should relate to the economic or financial standing, or to technical knowledge and ability of the company in relation to the contract in question. It is difficult to see how the factors concerning the ability of the company to take on long-term unemployed could be relevant to these factors. A possibility is that criteria for the award of most economically advantageous tender could take into account the employment of the long-term unemployed. The employing of long-term unemployed might bring about a compensatory saving in state expenditure which could be taken into account. If this was a criterion it would have to be amongst those listed in the relevant notice in the *Official Journal*.

Implementation of the directive

The directive came into force in July 1990 except in Greece, Spain and Portugal, where the implementaion may be delayed until 1 March 1992 as part of their general adaptation to EC requirements following their relatively recent joining of the Community.

PROCUREMENT BY THE UTILITIES

As noted above, the water, energy, and transport sectors were not covered by the supplies and works directives even when the awarding body was a public, or state owned entity. The supplies directive also excluded the telecommunications sector. These exclusions were because the pattern of ownership varied greatly between Member States and it was thought unjust to require different degrees of market opening. The Treaty articles on non-discrimination, such as Article 30, applied directly to public or state owned entities, even if their purchasing was not covered by the directives. The White Paper on the Completion of the Internal Market expressed the resolve to deal with this.

In October 1988 the Commission published three documents related to the utilities sector – a proposal for a directive on the procurement procedures of entities providing water, energy, and transport services [6.15], a proposal for a directive on the procurement procedures of entities operating in the telecommunications sector [6.16], and a 'communication' related to all these sectors [6.17]. The communication contained an analysis of the industries concerned, together with an assessment of government influence on purchasing behaviour and its consequences for the industries and their suppliers. The communication concluded that there was firm evidence that government influence resulted in a nationalistic pattern of procurement, even where government was not the owner of the industry. The consequences identified were higher costs for the public and fragmentation of the supplying industries, with adverse consequences for their competitiveness in world markets. The potential benefits of cost-saving and industry restructuring were considerable and warranted the imposition of a procurement regime similar to that applying to government procurement. To ensure uniform coverage and maximum impact the regime needed to be applied to public entities and to entities which were subject to special or exclusive rights, on the grounds that the granting of special or exclusive rights generally freed the entities concerned from full competition and was associated with a declared or tacit requirement for the industries to support national or local industry.

During subsequent stages the two proposals, which shared much in common, were merged into one, and a directive was finally published as 90/531/EEC of 17 September 1990 [6.18]. The discussion stages were marked by much lobbying and protest, particularly from the oil and gas sector, and the UK utilities, particularly newly privatised entities such as British Gas. Some of the major French utilities, such as Electricité de France, also protested, arguing that, because they had centralized purchasing systems, a much higher proportion of their purchasing would come above the thresholds of the directive than those of entities in decentralised countries such as Germany.

Another issue which the French in particular wanted linking with the opening up of procurement of supplies and works was the opening up of the markets in energy. This is a problem because France, which uses the highest proportion of nuclear generated energy in the world (70 per cent), has a surplus capacity which it wants to be able to sell in open competitive markets. Germany, on the other hand, produces relatively expensive electricity, mainly because there are agreements obliging the generators to buy expensive German coal. There is also reluctance to use nuclear energy on environmental grounds, a reluctance exacerbated by the closeness of several nuclear power plants to the German border. In witness to this unresolved problem, the directive contains the preamble:

> 'Whereas the Commission has announced that it will propose measures to remove obstacles to cross-frontier exchanges of electricity by 1992; whereas procurement rules of the type proposed for supplies of goods would not make it possible to overcome existing obstacles to the purchases of energy and fuels in the energy sector; whereas, as a result, it is not appropriate to include such purchases in the scope of this Directive, although it should be borne in mind that this exemption will be re-examined by the Council on the basis of a Commission report and Commission proposals'.

Application of the directive and aspects of the provisions

The directive applies to both goods, including any necessary siting and installation, and works. It also applies to software services procured by telecommunications entities and for use in 'the operation of a public telecommunications network or [where they] are intended to be used in a public telecommunications service as such'.

The most interesting aspect of the directive is that it applies requirements much like those in the works directive to a wide range of entities, many of whom are in the private sector and who may be used to operating internal procedures without outside involvemen..

The directive applies to public authorities, public undertakings (a body which is under the dominant influence or control of a public authority) and to other bodies in the relevant sectors when they 'operate on the basis of special or exclusive rights granted by a competent authority of a Member State'. Such special or exclusive rights are rights which have 'as their result the reservation for one or more entities of the exploitation of one of the defined activities.' Any body which 'may take advantage of a procedure for the expropriation or use of property or may place network equipment on, under or over the public highway' is amongst those considered to enjoy special or exclusive rights.

The activities covered are:

(1) 'The provision or operation of fixed networks intended to provide a service to the public in connection with the production, transport or distribution of drinking water, electricity, gas, heat' or supplying any of the above to such a network.

(2) The exploration and extraction of oil, gas, coal or other solid fuel or the provision of airport, maritime or inland port or other terminal facilities to carriers by air, sea, or inland waterway.
(3) Operation of transport networks including railway, automated systems, tramway, trolley bus, bus or cable. This only applies when the service to be provided is specified by the authorities. Thus the bus and coach operators in the UK would not at present come within the scope of the directive, while bus operators in many other countries would be covered.
(4) 'The provision or operation of public telecommunications networks or the provision of one or more telecommunications services'.

These sectors are conveniently referred to as the WETT activities, water, energy, transport and telecommunications.

When a private entity merely produces drinking water or electricity for its own needs and feeds its surplus into a network it is not covered, provided that the amount fed into the network has been less than 30 per cent of total production over the last three years. If the gas or heat is an unavoidable by-product, revenue from its sale may be up to 20 per cent of the entity's turnover on average over the past three years, before the entity is covered.

The appendices to the directive list various entities in each country which are covered, though the list is not necessarily exhaustive. In the UK these include the water companies, British Gas plc, entities granted exploration rights under the Petroleum (Production) Act 1934 as extended by the Continental Shelf Act 1964, British Coal, British Rail, Greater Manchester Transit Company, Docklands Light Railway, London Underground, Tyne and Wear Metro, airports managed by British Airports Authority plc, Harbour Authorities within the meaning of section 57 of the Harbours Act 1964, British Telecommunications plc, Mercury Communications Ltd, Racal Vodafone and Telecoms Securicor Cellular Radio Ltd (Cellnet).

The purchase of water by the water providers is not covered by the directive, which allows transfers of water between nearby entities to take place without being subject to the rules. Purchases of energy and fuels by the entities supplying electricity, gas or heat, in oil, gas or solid fuel exploration and production are not covered. This avoids the problem of there being no free market in energy for the time being.

Alternative regime for oil and gas exploration and production

To meet complaints that the normal rules would impose significant extra costs and delays on companies working in oil and gas exploration and production, and that in any case the industry was so competitive that the directive was unnecessary, a special exclusion, which provides in effect an alternative regime, was devised. Under this regime Member States can apply to the Commission to have the sector exempted in relation to activities in their territories. Exemption is only granted if certain conditions are met. These relate to freedom of companies to enter the sectors, non-discriminatory conditions and practices, and the absence of any requirement on companies to provide information about procurement (those holding concessions in the UK sector of the North Sea have been required to provide such information to the Offshore Supplies Office). None of these conditions relate to standards, which was an area of major concern to the companies who tend to use their own and American Petroleum Institute standards and who consider that existing standards at national or community level in the EC are inadequate. However, it is not at all certain that the UK, the Member State with approximately 90 per cent of EC production in the sector, intends to utilise this exemption.

Some of the differences in relation to the supplies and works directives

More freedom about the use of procedures and standards is the main difference. For example, entities may derogate from the requirement to use European specifications where 'the relevant European specification is inappropriate for the particular application or does not take account of technical developments which have come about since adoption'; the reasons for the claimed inadequacy have to be communicated to the appropriate standardisation body.

'Framework' agreements may be used. These are agreements which establish terms, including prices, for a possibly unspecified quantity of procurement over a given period. The use of such agreements, if let following the requirements of the directive, free the entity from need to advertise the individual purchases under the agreement.

Entities have freedom to use the open, restricted and negotiated procedures provided that, in most cases, a call for competition has been made. Periodic indicative notices are to be issued at least once a year, but for supply contracts, only the total projected purchasing in each product area needs to be listed, and then only if it is likely to exceed ECU 750 000 during the year. Entities can also avoid the need to advertise each contract by operating a qualification system, provided that an advertisement about the existence of the system is placed at least once a year (unless the system is to operate for less than three years).

There is provision to enable advantage to be taken of special bargain offers by suppliers.

One controversial provision obliges undertakings to accept supplies originating from an EC country (plus other countries with which agreements have been entered into) provided that the price is not more than 3 per cent higher. Tenders for supplies may also be rejected where the third country content exceeds 50 per cent. Complicated rules of origin, defined in regulations, are applicable. These provisions caused alarm amongst the free traders, both inside the EC, for example the UK government, and outside, with talk about 'fortress Europe'. The Commission argued that the provisions were there as a bargaining counter to help prise open external markets. The USA and Japan are the countries such a provision would have been aimed at.

For works the threshold, at ECU 5 million, is the same as that in the works directive, while the threshold for supplies is ECU 400 000, except in the telecommunications sector where it is 600 000 ECU (cf. ECU 200 000 threshold in the supplies directive).

Implementation

The directive is to be transposed into national law by 1 July 1992, but the measures adopted by Member States may take effect six months later on 1 January 1993. Spain may delay the coming into force of the measures until 1 January 1996, while Greece and Portugal can delay it until 1 January 1998.

Proposed services directive

The proposal for a directive on the co-ordination of procedures on the award of public service contracts was issued in December 1990 [6.19] and if adopted will open up procurement above the thresholds for all design and consultancy services relating to the construction industry, amongst many other services. The proposal faces much opposition and may be reduced in scope. Some opening up of the markets is almost certain, particularly in construction related services. Work is proceeding on a similar proposal covering the procurement of services in the utilities sector, but the shape of that proposal will be strongly influenced by what happens to the public procurement of services proposal.

Aspects of the proposal

Perhaps surprisingly, the structure and general wording is very close to other public procurement directives. All the elements, such as advance notices, provisions on technical specifications, choice of award procedures, calls for competition, criteria for selection of tenderers or participants, and criteria for the award of contracts are present, with occasional changes and additions as necessary. Most of the supplementary provisions, such as those relating to employment of sub-contractors and providing information on employment conditions to contractors, come straight from the works directive. However, the implications of the directive are very great.

The proposed directive would apply to the same authorities currently subject to the supplies directive and the works directive. It covers contracts for all types of services, but these are divided into two categories: Annex 1A services, within which services associated with construction projects are listed, and which are subject to the full provisions of the directive; and Annex 1B services, which are subject to very limited provisions covering the use of European standards and post-award notices and statistics to be sent to the Commission and not necessarily published. Annex 1A includes some transport services, certain financial services such as insurance, computer and related services, management consulting, architectural services, engineering services, urban planning and landscape architectural services, related scientific and technical consulting services, technical testing and analysis services, building cleaning services, sewage and refuse disposal, sanitation and similar services. The Annex 1B list includes legal services, educational and vocational services, health and social services, and recreational, cultural and sporting services, amongst others.

Where a public authority is responsible for providing a service, and wishes to have another entity carry out the work with entitlement to charge customers for providing it, the award of the concession is subject to the provisions of the proposed directive. When concessionaires purchase services they are to be subject to the rules on advertising contracts above the threshold value.

Work carried out by one authority on behalf of another, such as the contract management services offered by the Property Services Agency in the UK, must be opened up to competition unless there is an exclusive right according to a published law, regulation or administrative provision which is compatible with the Treaty. This could have had considerable implications in the UK if agencies had not already been privatised or subject to competition, and might have much impact in certain EC countries where much design and consultancy work is done by government agencies.

The proposed threshold is ECU 200 000. The original proposal did not apply this threshold to the design of public works as defined in the public works directive. Instead it was proposed that where the estimated cost of the public works was not less than ECU 5 000 000 (i.e. the threshold in the public works directive) design work in connection with it should be subject to the public services directive. There were also complicated rules, such that where the public works was divided into lots design in connection with lots of less than ECU 1 million were not to be subject to the rules of the directive, provided that the design of not more than 20 per cent of the estimated overall value of the work was exempted. Apart from some problems of wording the result would have been that some exceedingly small design contracts could have been subject to provisions of the directive, particularly since design work is in any case commonly sub-divided.

The proposal makes provision for design contests, which are commonly used in some EC countries, particularly Germany, for architectural work. Where prizes or payments are not less than ECU 200 000 there are requirements designed to ensure impartial adjudication and that no geographical limit (within the EC) on entries to the contest is imposed.

The provisions on technical specifications would have the effect that design would have to be carried out to European standards and codes, and if contract documents are to be prepared, their format could be prescribed by European standards, if and when standards are adopted in this area.

On the evaluation of participants or bidders the proposal introduces the possibility of requiring a check by a competent official body of the country in which the supplier is established, subject to that body's agreement, on the technical capacities of the supplier and, if necessary, on his study and research facilities and quality control measures. There is also reference to a description of the suppliers' measures for ensuring quality, and his study and research facilities. The subject of quality assurance certificates is also introduced. Certificates from equivalent bodies in other Member States must be accepted and 'other evidence of equivalent quality assurance measures from suppliers who have no access to such certificates, or no possibility of obtaining them within the relevant time limits' are to be accepted. Clearly the Commission has recognised the developments taking place in some Member States, the UK particularly, in the field of externally awarded quality assurance certificates, and the fact that in some Member States the introduction of certification systems lags behind.

Another provision without equivalent in the works directive clearly attempts to tackle the problem of public services organisations, for example in-house design teams, benefiting in some way from their protected position and providing unfair competition against the private sector. 'In the case of a tender submitted by a public authority or a public enterprise, the authority shall in particular enquire whether the tender is influenced by public funds allocated' in various ways, and 'shall inform the Commission if it intends to award a contract to a tenderer whose tender is influenced by' the award of public funds in the ways described. However, the problem of in-house teams providing services such as design, consultancy and project management to a degree which varies widely among Member States has not, so far, been tackled. The bodies representing consulting engineers would particularly like some intervention in this area.

Provided the proposal is adopted, and in a form not too weakened, the public services directive could have a significant impact in an area where there is considerable scope for opening up of markets.

Enforcement

One of the keystones in the strategy of opening up public procurement is to improve observance of the directives and the Treaty through active enforcement. The Commission has done this by setting up a section within DGIII which monitors notices in the *Official Journal* and which responds to complaints and various indications of non-compliance. With the new directives bringing about important changes in the procedures and notices regarding supplies and works, the volume of work and the possibility of detecting non-compliance has sharply increased. Another approach has been to check the procedures when European Community structural funds have been used to support a project. It is now routine for the release of the funds to be made conditional on a check by the section to ensure that the public procurement directives were complied with. Part of the work of the section is to provide an information service so that the authorities awarding contracts are clear as to the requirements.

One of the problems under the old enforcement system was that when an authority was in breach, the Commission was only empowered to launch an action against the Member State, in accordance with Article 169. This starts with a *mise en demeure* (formal notice) followed by a 'reasoned opinion' (sometimes referred to using the French term *avis motivé*) delivered to the Member State. The collecting of information about the alleged infringement would normally be dealt with through the office of the Permanent Representative of the Member State in Brussels. If the Member State does not comply with the opinion within the period laid down by the Commission, the latter may bring the alleged infringement before the Court of Justice. In practice, the penalty is mainly political because the court does not have the power to impose penalties in such instances.

The system was not without power, and many cases have been dealt with quite effectively. One of the better reported cases involved the Danish government and main construction contract for the Storebælt Western Bridge. The contract conditions obliged contractors to use, to the greatest possible extent, Danish materials, goods, labour and equipment, but the action mainly concerned the conduct of the Danish authorities during the tendering period. Particularly, one of the firms, having submitted a tender, was apparently told during a tender 'clarification' meeting that the Danish content needed to be increased. It was also alleged that the successful contractor, who had submitted an alternative tender, had been involved in negotiations contrary to the procedure for restricted tendering, and that the alternative design had been successful despite the fact that the conditions accepted were contrary to those stipulated. In particular, any contractor submitting an alternative offer was, according to the tender documents, to be fully responsible for the design risk and any changes in quantities associated with the alternative design. These conditions had been waived.

A complaint to the Commission from one of the joint ventures whose tender was rejected resulted in very rapid action involving the Commissioner for the Internal Market, Mr Bangemann, and the Danish government. At one stage, court action was imminent and there was the threat of interim measures applied by the court, possibly involving temporary suspension of the award which could have caused substantial problems for the whole project. In the event the matter was settled out of court and the Danish government agreed that 'unsuccessful bidders for the bridge contract would be able to reclaim the costs of their tenders and possibly seek compensation for loss of profit, subject to proof that losses were caused by illegal clauses in tender documents' [6.20].

A new procedure

It was clear that improved enforcement procedures were needed, particularly in certain Member States, including the UK, where there was little possibility of legal remedy if the authorities failed to follow the provisions of the Treaty and the directives in their award procedures. The 'remedies' or 'compliance' directive, properly called the 'Council Directive of 21 December 1989 on the co-ordination of the laws, regulations and administrative provisions relating to the application of review procedures to the award of public supply and public works contracts' [6.21] was the response to this. Because the provisions had such important implications throughout the widely differing judicial institutions and traditions of the Member States, the directive involved painstaking research and discussion with authorities throughout the EC.

The provisions are relatively brief. The main requirements are that Member States should set up independent judicial bodies with power to suspend the award of contracts and to set aside any decisions which are in breach of the public procurement directives, national law implementing them or the Treaty. The body must also have the power to award damages to persons harmed by the infringement.

On the subject of damages claimed there is little guidance in the directive itself, but the wording of the proposed remedies directive [6.22] covering procurement in the utilities (formerly excluded) sectors may clarify the intention:

> 'Where a claim is made for damages representing the costs of preparing a bid or of participating in an award procedure, the person making the claim shall be required to prove the infringement of Community law in the field of procurement or national rules implementing that law and that the infringement adversely affected his chance of being awarded the contract. He shall not be required to prove that, in the absence of the infringement, he would have been awarded the contract. The amount of such costs shall be deemed to be 1 per cent of the value of the contract unless the person making the claim proves that his costs were greater'.

Originally the Commission wanted the power directly to suspend an award procedure, but there was such opposition to this from the Member States that an alternative was devised. The directive provides a procedure, similar to the Article 169 procedure of the Treaty, for the Commission to draw the attention of the authorities in the Member State to the infringement, so that rapid correction can take place. If the Commission is not satisfied with the response of the authority in the Member State, the Court of Justice has agreed to make prompt interim rulings, which could include the suspension of the award procedure.

UK response

The response to the directive in the UK was negative, characteristically negative some might say. A representative of the Institution of Civil Engineers said the directive 'bears no relation to reality . . . giving disgruntled tenderers the opportunity to hold up the tender process would make the whole system unworkable' [6.23]. Mr Colin Middlemiss of the Building Employers Confederation said [6.24]:

> 'Our own view is that it won't encourage UK contractors to rush overseas and start tendering for jobs. The risks in cross-border trading are in other directions. Margins are not great, building regulations differ and there are industrial relations and supply problems . . . This [the power to suspend award procedures] will make no difference, it's already possible in the UK for the contractor to go to a court and gain an injunction to stop an award.'

However, he admitted that the directive does 'give teeth' to the enforcement of fair play in the post-award stage: 'If a contractor thinks he has been excluded from a tender list for discriminatory reasons he would be able to complain to the judicial body and could receive damages.'

Mark Lewis, a senior associate with a firm of London solicitors, writing in the *Financial Times*, commented on the legal implications of the directive:

> 'In English law sanctity of contract is still pervasive. The categories of contract liable to be set aside by the operation of legal doctrines, such as illegality or public policy, are relatively limited. Any addition to these categories will be followed with interest by lawyers, some of whom will surely benefit from the inevitable litigation that will follow from applying these procedures.' [6.25]

The threat of claims for damages, and an army of watchful contractors, with recourse to the Commission, seeking to benefit from lapses by the authorities is likely to have the effect of dramatically improving compliance with the rules.

Plans for future directives in the field of public procurement

As referred to above, plans for future directives are primarily in the field of extending rules in the utilities sectors, firstly to the procurement of services, and secondly to providing for effective enforcement and remedies.

Possible development of the GATT government procurement code

One issue, held up at the time of writing by the failure of the Uruguay round of the GATT agreement, concerns the extension and broadening of the GATT code on government purchasing. As discussed above, the present code is restricted to supplies (above a threshold of SDR 130 000 (ECU 134 000) purchased by central government. The main objectives of the negotiations have been to extend the code to services, including construction, to extend it to local government procurement, and possibly most important of all for EC industry as a whole to cover procurement by the water, energy, transport and telecommunications industries (as the utilities directive does within the EC). The EC has undertaken studies which indicate competitive advantage in some of these industries, but none more so than in railway equipment, particularly urban transport systems, where at present the USA operates price preferences in favour of US products of up to 25 per cent.

However, the USA claims that its federal government does not have the power to impose purchasing conditions on state authorities, let alone regional authorities, and furthermore, that purchasing by the utilities is not subject to political influence. The studies carried out by the EC indicate considerable influence. Japan also claims that its utilities are not subject to pressures to buy national and that it would be unable to impose procurement conditions on them. Regarding construction in Japan, both the USA and the EC have been applying pressure to get the Japanese to open up their public contracts, which have proved extremely difficult to gain access to. These markets are large because the country has much infrastructure to develop to get to Western standards and the pace of economic growth has been sustained and rapid. The country also suffers from a shortage of labour, particularly construction labour, and Western companies believe that they could make a substantial contribution to the development of the country. The USA, in addition to their efforts in the GATT Uruguay round negotiations and the GATT government procurement negotiations, has undertaken bilateral negotiations to prise open the market in construction. In one agreement, the Japanese government agreed to give special access for USA companies to 14 mega projects such as the construction of Kansai airport. Even this initiative met with limited success with the Japanese authorities rejecting bids on the grounds that included lack of experience with the particular characteristics of Japanese earthquakes. Other problems have been related to the licensing system for construction companies and to the 'dango' system of contract allocation, where contracts are distributed on a semi-official rota basis to local firms.

The opportunities under a revised GATT government procurement code could be considerable. In any case, the Commission is anxious to give whatever help it can with overcoming institutional obstructions encountered by companies. The total volume of purchasing by government and publicly controlled agencies around the world is estimated to be worth up to $1700 billion a year, with a high proportion accounted for by Japan, North America and, of course, the EC. The outcome of the negotiations, which are closely tied to the main GATT negotiations, should be of considerable interest to industry, including the construction industry.

Public procurement and regional and social matters

The Commission has for some time wrestled with the problem of how to dismantle the existing schemes of regional preference in force in the EC, notably in the former West Germany (FDR), Greece, Italy (the 30 per cent Mezzogiorno quota, as referred to above, was in 1990 ruled to be contrary to the Treaty) and in the UK. These schemes have provided mechanisms for using public purchasing to support firms, usually of small or medium size, in less developed regions or in regions of high unemployment.

The Commission produced a communication on the subject of public procurement and regional and social aspects [6.26] released in 1989. The Commission expressed the aim 'to undertake urgent discussions with Member States concerned with a view to bringing any infringements to an end at the latest by 31 December 1992'. With regard to the Mezzogiorno system, the court seems to have done the Commission's task for them. The communication indicated that the impact of the schemes was very small. In the former West Germany the regions which benefited accounted for about 15 per cent of the population, but the value of contracts affected was only 0.25 per cent of total federal procurement. In the UK, preferences, which were in any case administrative rather than legal in form, were applied to contracts worth only 0.02 per cent of total public procurement and the total additional cost of operating the schemes was a miserly ECU 17 000. In Greece the schemes covered 90 per cent of the area of the country, containing 65 per cent of the population, but because of the lack of suitably qualified Greek firms for many of the contracts the import penetration was 40 to 45 per cent of public procurement, which is much higher than in any other Community country.

One of the problems is how to target a scheme to provide the necessary support without breaching the Treaty provisions providing for non-discrimination. The communication considered the possibility of various types of socially motivated conditions, such as the one referred to in the Gebroeders Beentjes case discussed above. It concluded that the use of such conditions was possible without infringing the provisions of the Treaty, but that conditions would have to be considered virtually on a case by case basis to establish whether they could have a discriminatory impact on non-national firms. For example, 'a condition that firms employ a given number or proportion of trainees exclusively from national training programmes would appear to discriminate against trainees in equivalent schemes in other Member States. It would accordingly not be admissible. On the other hand, an obligation to employ trainees which does not limit their provenance on national lines would not appear to pose a problem'.

Much of the communication is devoted to considering schemes for helping small and medium-sized enterprises, and particularly to help them win public contracts. Methods proposed included training, sub-division of contracts, encouraging sub-contracting. The general question of assistance for small and medium enterprises is discussed in chapter 5.

The communication leads one to conclude that regional preferences offer little scope and that the problems of regional disparities should be tackled by other initiatives linked to the European Regional Development Fund, the European Social Fund and the European Investment Bank.

The position at the time of writing is that the UK and German schemes referred to above are defunct. The others, and a new German scheme related to the former East Germany, are the subject of infringement procedures.

Chapter 7

Social Policy, Employment Conditions and Health and Safety

SOCIAL POLICY

The single market is concerned with removing barriers to competition. Many less competitive firms will disappear, particularly in the realm of transportable services, such as insurance, and manufacturing. Estimates suggest between two and five million net additional jobs resulting from the growth produced by removing barriers. But the restructuring of industry will cause many people to lose jobs, be retrained or even move. For the market to operate efficiently, particularly in times of technological development, resources, including labour, must be free to move from one industry to another. Old skills become redundant and new skills have to be acquired. The workforce needs to be both adaptable and physically mobile, within each Member State, and possibly between Member States. With change and transition come stress, and employee resistance. The numbers of long-term unemployed may increase with many unwilling or incapable of learning new skills or of moving to the opportunities.

The British 'right' believe that the benefits of prosperity percolate naturally to all sectors of society with government's prime responsibility the removal of the barriers to growth. However, most EC governments, supported by the Commission and particularly its President, Mr Delors, believe that market mechanisms alone cannot be relied on. Liberalisation of markets must be accompanied by parallel measures to safeguard the interests of weaker groups and less developed regions. This is seen as an end itself, but also to make the single market politically acceptable, and to bring about 'social and economic cohesion'.

The social dimension is intended to counterbalance the adverse effects of competition. To UK sensibilities, accustomed to free market policies being associated with the right, markedly so under Mrs Thatcher, and 'social policies' being associated with the left, the EC mix of market liberalisation and social measures seems uneasy. The UK agreed to the Single European Act but tends to treat the resulting proposals like an *à la carte* menu. Mrs Thatcher denounced the Social Charter (see opposite), which embodies the social elements, as Marxist interventionism and rejected any attempt to regulate working conditions, other than those involving health and safety. The British trades unions, meanwhile, see the social dimension as an opportunity but tend to view aspects of market opening and industrial restructuring as a threat to be fought.

The construction industry will be affected as employers, whether at home or elsewhere in the Community. Community policy on health and safety is very active and will have a major impact on construction sites and, significantly, in the area of project planning where new ground is likely to be broken by the proposed directive on temporary or mobile sites. The emphasis in recent legislation and proposals in the area of health and safety is very much on assessment, planning and risk avoidance. Apart from changes in employment and occupational health and safety areas resulting from Community initiatives, those in the construction industry should be aware of the considerable differences in philosophy, existing legislation and enforcement around the Community in these areas.

The social dimension and the Single European Act

Concern about the social consequences of competitive changes brought about by the Single European Act resulted in it including a new section on Social Policy (Title III of the Treaty, Articles 117 to 128). This introduced into the Treaty new Articles, 118a and 118b, and a complete new Title 'Social and Economic Cohesion' with another five new Articles, 130a to 130e. Social and economic cohesion is not defined, but the harmonious development of the Community is referred to and the term is generally used in connection with reducing differences between regions, social groups and making provision for long-term unemployed. The various structural Funds (the European Agricultural Guidance and Guarantee Fund, Guidance Section, European Social Fund, European Regional Development Fund) and the European Investment Bank are to be used to these ends.

Article 118a is concerned with 'improvements, especially in the working environment, as regards the health and safety of workers', and has the objective of 'harmonisation of conditions in this area'. Qualified majority voting applies to measures introduced under this article. Article 118b requires the Commission to 'endeavour to develop the dialogue between management and labour at European level which could, if the two sides consider it desirable, lead to relations based on agreement.'

What is the social dimension?

The remit of the European Parliament's committee on social affairs, employment and working environment indicates the scope of the social dimension:

(1) Living and working conditions including paid holiday arrangements.
(2) Health and safety at work.
(3) Employment policy especially in relation to disabled and young people.
(4) Wages, pensions, incomes and capital formation.
(5) Vocational training and vocational qualifications.
(6) Activities of the European Social Fund.
(7) Free movement of workers.
(8) Social security of all migrant workers and their rights.
(9) Housing policy.
(10) Equal treatment between the sexes.

The social dimension also includes public health, for example, smoking and AIDS, and there is even to be a Community wide telephone emergency number in the interests of public safety. The number is to be 112, though this number can remain in parallel with existing emergency numbers until 1995.

THE SOCIAL CHARTER

An article in the *Financial Times* of 12 June 1989 was entitled 'Storm cloud gathers over the Social Charter' and commented that it had assumed disproportionate symbolic status in the eyes of both protagonists and antagonists. Mrs Thatcher denounced it while the Commission warned that 'mounting popular discontent could scupper the whole internal market project unless workers are given an adequate safety net'. The 'Community Charter of the Fundamental Social Rights of Workers' was finally agreed by all the Heads of State or Government of Member States, except the UK, on 9 December 1989. It is no more than a declaration, with a list of 26 rights and four articles on the implementation of the charter. Member States are to guarantee the rights and the Commission is invited to put forward initiatives with a view to adoption of legal instruments. The Commission is also required to produce an annual report on the application of the charter. The Social Charter is important because it indicates the likely content of future proposals relating to employment and health and safety. Many recent proposals clearly tie in with the rights contained in it.

Rights under the Social Charter

Freedom of movement of workers is guaranteed and is deemed to imply harmonisation of conditions of residence, elimination of obstacles arising from non-recognition of qualifications. All employment is be fairly remunerated and workers should be 'assured of an equitable wage, i.e. a wage sufficient to enable them to have a decent standard of living'. There should be an 'equitable reference wage' for workers not in 'open-ended full-time' employment.

Living and working conditions are to be improved with the completion of the internal market. Reference is made to approximation (i.e. converging of differing national practices), while improvement is maintained 'as regards in particular the duration and organisation of working time and fixed-term, part-time, temporary and seasonal work'. Workers are entitled to weekly rest periods and annual paid leave. Every worker is entitled to social protection and an adequate level of social security benefits. Those who cannot enter or re-enter the labour market should have 'sufficient resources and social assistance in keeping with their particular situation'.

The broad right to strike is one of the rights, but is subject to the national need to exclude sections of the armed forces, the police and civil services. Employers and workers have the right to join or not to join associations without 'personal or occupational damage'. Dialogue must be developed.

Training and retraining facilities should be provided. There should be equal treatment between the sexes. There should be information provided, presumably about the firm and management issues, and consultation and participation for workers should be developed along 'appropriate lines', and particularly in companies or groups which operate in more than one Member State.

Health protection and safety at the workplace should be improved and harmonised. Training, information, consultation and balanced participation of workers are stressed in this context.

Children and adolescents should be protected and provided with vocational training (during working hours). The minimum school leaving age should be not less than 15. Elderly persons should be provided for, and disabled persons given special assistance.

One provision *not* in the Social Charter concerns sub-contracting using cheaper labour from, or even in, another Member State, for example to provide components or to carry out assembly work. The preliminary draft [7.1] contained a clause 'the wage conditions as well as other social benefits relating to this wage applied in the host country must be in particular guaranteed to workers of another European Community Member State performing work for the account of a sub-contracting undertaking in the host country.' This was to satisfy concerns, particularly from the left in the European Parliament, that the internal market would result in undercutting of workers in northern countries by cheaper labour with lower social overheads. The intention, if translated into legislation, would have undermined much of the competitive advantage of the less developed countries in the Community and ruined their chances of rapid catching-up. It would also have been a barrier to carrying out cross-border construction work using sub-contractors from outside that Member State.

Action programme [7.2]

The Commission action programme of November 1989 relating to the implementation of the Social Charter listed supporting initiatives, both proposed and existing. The Commission acknowledged that it had no role in wage-setting but it stated that it had a responsibility to issue an opinion in relation to establishing a 'decent reference wage' and the 'maintenance of remuneration in respect of periods of leave or sickness'. Interestingly, in view of later arguments (see p 124), the proposal for a directive on 'the adaptation of working time' is listed under 'improvement of living and working conditions' whereas when the proposal was issued the legal base chosen was Article 118a, i.e. health and safety, which only requires a qualified majority for adoption. One of the proposals in the action programme is for measures to improve the access of disabled people to transport, which should affect public transport planning and costing.

EMPLOYMENT CONDITIONS

Impact of existing Community legislation on employment law in the UK

The general impact of Community legislation on employment law in the UK has been limited, despite a considerable number of directives and recommendations in the area. Possibly the area most affected is that of sexual equality, but there have been other changes, for example in the areas of employee protection in cases of collective redundancy when undertakings change hands, or go into liquidation. The case of *Litster* v. *Forth Dry Dock Co Ltd* [7.3] discussed in chapter 4 concerned the interpretation of the regulations which implemented Council Directive 77/187/EEC. In that case the employer tried to circumvent the regulations which required an undertaking taking over another, to be responsible for continuing the former employer's contracts of employment. By dismissing the employee and then re-employing an employee it was hoped to avoid the responsibility. However, the judge held that where the regulation in question was required to implement EC legislation it was necessary to put a purposive construction on the words, following the practice of the European Court. On this basis the obligation was not avoided by the technicality of the break in employment.

Sex discrimination

There have been a number of directives in this area. Directive 75/117/EEC required the abolition of laws contrary to the principle in Article 119 of the Treaty that 'men and women should receive equal pay for equal work'. In effect the directive provides elaboration of the article by requiring that '. . . for the same work or for work to which equal value is attributed, the elimination of all discrimination on grounds of sex with regard to all aspects and conditions of remuneration'. Other directives have related to access to employment, vocational training, promotion, working conditions, social security schemes and treatment in self-employment. In a recent proposal for a directive even apparently neutral provisions may be held to be discriminatory if a 'criterion or practice disproportionately disadvantages the members of one sex' and is not objectively justified.

However, it is in the case law on sex discrimination that the effect of Community law has been most noticeable. In *Marshall* v. *Southampton and South West Hampshire Area Health Authority* [7.4] Miss Marshall was dismissed on reaching her sixtieth birthday. Because this was a 'provision in relation to retirement' it was permitted under the Sex Discrimination Act 1975, but was contrary to Council Directive 76/207/EEC, which had not then been implemented. It was held that she could rely on the directive even as an individual, and that it could have effect against an emanation of a Member State even when it was acting merely as an employer. In *Pickstone* v. *Freemans* [7.5], five women warehouse operatives claimed that their work was of equal value to that of a male warehouse checker. The court felt it necessary to give a purposive interpretation to the regulation which implemented

Directive 75/117/EEC. Using this method of interpretation it was possible to make the comparison, even though it contradicted the strict meaning of the UK law. Because in this case the employer was not an emanation of the state, it had not been possible to apply the directive directly as in the *Marshall* v. *Southampton etc.* case referred to above.

There have been several recent cases involving sex discrimination in pension arrangements. In the influential *Barber* v. *Guardian Royal Exchange Assurance Group* [7.6] the court considered a pension scheme in which the normal pensionable age was 62 for men and 57 for women. If made redundant an immediate pension was available for men at 55 and for women at 50. The court had to consider whether Article 119 of the treaty, which provides for equal pay for equal work, applied to pension arrangements. It was held that all aspects of the remuneration package granted to employees were within the scope of the article.

If the decision had applied to existing pension schemes retrospectively, it could have threatened the solvency of many of them. The court was careful to explain that pension rights to the date of the judgment would not be affected. The judgment overturned the widespread view, held by the Commission of the EC amongst others, that Article 119 did not affect pension ages. Private UK pension schemes have generally responded by adopting the same pensionable age for men and women for new policies.

Depending on the outcome of a recent challenge in the European Court there may soon be equal treatment even in that bone of contention for men, the official retirement age. On 3 December 1990 the High Court in London decided to refer the operation of the state pension scheme to the European Court as to whether the differential retirement ages for men and women is contrary to EC law.

Proposed legislation concerned with employment conditions

There are five new proposals in the pipeline or under reconsideration at the time of writing. Three are concerned with part-time and temporary workers, but have been rejected by the Council of Ministers. They are likely to re-emerge in amended form. Another proposal is concerned with working hours and the final one with maternity leave.

The proposal about benefits for part-time, temporary and intermittent workers and on working hours have aroused strong feelings in the UK, both in government and industry. The consultative document issued by the Department of Employment on 3 August 1990 put forward detailed arguments against the proposals and explained that

Table 7.1 EC figures for changes in persons employed 1979–88 (in millions) [7.7]

	Total employment	Full-time equivalents
Germany	0.95	0.7
Italy	0.7	0.7
Netherlands	1.0	0.3
France	-0.2	-0.4
UK	0.5	-0.52

the philosophy behind the directives ran counter to the government's own policy of 'reduction of unnecessary burdens on employers and the removal of disincentives to recruit new employees.' It went on to claim that the policy had had considerable success. 'Since March 1983 the UK has experienced the longest and most productive period of job creation since the Second World War. Over 3.5 million additional jobs have been created in seven years. Over this period some 1.3 million part-time jobs have been created. There are now some 5.5 million part-time and temporary employees, representing about 25 per cent of the employed labour force.' Figures for the changes in numbers of persons employed between 1979 and 1988 quoted by the Commission (see Table 7.1) convey a very different impression (clearly the different base date is crucial).

Ministers claimed that full implementation of the directives would cost £3 billion and accused the Commission of abusing the Treaty of Rome by using 'inappropriate articles'. Mr Tony Blair, shadow employment secretary, meanwhile, said that this was ludicrous nonsense and that the figures had merely been plucked out of the air.

Conditions for part-time, temporary and intermittent workers

The proposal (COM (90) 228, 26 June 1990) for conditions for part-time, temporary and intermittent workers (part-time, etc. workers) was based on Article 100 of the treaty, which requires unanimity. It would only have affected those working more than eight hours a week. The following provisions were included in relation to part-time and temporary or intermittent work:

(1) Compulsory access to vocational training comparable to those in full-time work.
(2) Where national requirements for worker representation and

consultation are based on numbers of workers in a firm, part-time and temporary workers would have been counted pro-rata.

(3) The obligation to inform workers' representative bodies before part-time etc. workers were used.
(4) Firms with over 1000 employees would have had to provide a regular report on the use of part-time etc. workers.
(5) Employers would have to provide part-time etc., workers with 'the same treatment as regular employees in respect to cash benefits and benefits under social assistance schemes and non-contributory social security schemes'.
(6) The same access for part-time etc. workers within undertakings to social services enjoyed by regular employees.
(7) The banning of contractual clauses aiming to prevent such workers from being recruited to full employee status.

Proposed directive on benefits available to part-time, temporary and intermittent workers

The directive on benefits for part-time etc. workers was published in the same 'COMdoc' as the above proposal. It proposed:

(1) That such workers should be given social protection under statutory and occupational social security schemes, account being taken of hours worked, as regular employees.
(2) Entitlement to annual holidays, dismissal allowances and seniority allowances, in proportion to hours worked.
(3) An overall limit on the time for which a temporary worker can be employed as such. The limit proposed was three years, but only applied where the periods of temporary work lasted 12 months or less. Also, where there were 'unjustified' gaps in between fixed periods of temporary work, the employee was to be entitled to an 'equitable allowance'.

The directive was to be applied to seasonal workers so far as 'the special features of this form of work allow'.

Because differences in national treatment in this area result in differing costs for employers and can thus distort competition between companies based in different Member States the legal basis chosen was Article 100a, the internal market article, despite the proviso in that article that it should not apply to the 'rights and interests of employed persons'. The UK objected to this legal base because it only requires qualified majority voting.

Progress in the Council of Ministers

The UK government argued strongly that the labour markets must be flexible and should not impose unnecessary costs on firms. It feared that it would be alone in objecting and that the 100a proposal would be accepted on a qualified majority. In the event, other states joined the UK. Germany voted against, fearing incompatibility with its existing social security system. The meeting of the Council on 26 November 1990 rejected the proposals as drafted and indicated that much modification was needed.

Health and safety of part-time, temporary and intermittent workers

This proposal concerns the health and safety of the part-time etc. workers referred to in the above proposals. The user of such labour is to be made responsible for applying health and safety provisions, even where the workers are employed by an organisation supplying temporary labour, but without lessening the liability of that organisation. Information and, where applicable, training are to be provided. Also, temporary workers should not be used for work where special medical supervision over a long period is required except in special circumstances. The latter provision would presumably apply to those breathing air under pressure, for example in tunnelling and diving, and to those working with asbestos or in the nuclear industries. The proposal was based uncontentiously on Article 118a, the health and safety of workers article.

Proposal for a directive on working time

The proposal on working time covers times when the employee (it is not to apply to the self-employed) is at the disposal of the employer, but not if on-call but not at the workplace. Night work is work taking place during a seven consecutive hour period, as designated nationally, between 2000 and 0900. Anyone who regularly worked three hours night work would count as a night worker. All employees would have to have at least 11 hours rest period in every 24 hours, and one 24 hour rest day every week averaged over not more than 14 days. The weekly rest period would have to follow a daily rest period giving at least 35 hours once a week. Minimum paid holidays would also have to be provided. Exceptions would apply in case of force majeure, in case of actual or potential accident, or in case of any seasonal or other special characteristics of the work or other 'exceptional situation'. If any of the exceptions were used compensatory rest periods would be required. Oil rig work is amongst those categories to which the exceptions would apply. The effect would be to restrict night shift work to eight hours.

A press release about the proposal referred to a link between long hours of work on construction sites and the high rate of deaths and injuries in the industry, but construction employers were particularly critical, referring to the need to make maximum use of daylight hours, to established methods of working in the tunnelling industry and in connection with lane rental contracts and railway possessions [7.8].

The Commission has based this proposal on Article 118A, which deals with health and safety at work, one presumes to take advantage of qualified majority voting in the Council. If it had been based on Article 100, dealing with harmonisation in general, unanimous voting would have been needed.

National differences in the regulation of conditions of employment including working hours

The idea of regulation of working hours is alien to the UK except where great risk to the public is concerned, such as for pilots and drivers of coaches and other commercial vehicles. However, every other EC Member State has legislation governing the working week, maximum overtime, or annual paid leave. Belgium, Germany, Greece, Spain, France, Ireland, Luxembourg, Netherlands and Portugal have legislation in all of the areas. In Germany, for example, the statutory limit on the working week is 48 hours, with up to two hours overtime per day for up to 30 days a year; 18 days annual paid leave are provided for, in addition to 10 to 14 public holidays.

The UK and Italy are the only Member States with no general statutory limits on overtime. Arguments are widely voiced in the UK about regulations bringing inflexibility. Concern is also expressed in other countries, particularly Germany, though this is muted because of fears that potential investors may be deterred.

Proposal for a directive on worker consultation

Another draft of a proposal for a directive issued by the Commission at the end of 1990 would require companies with over 1000 employees and operating in more than one EC Member State to set up a council of workers' representatives. An employer would have to report to the council on the company's affairs and consult on a range of issues including organisational changes and new working practices.

Again the British government is likely to veto this proposal, even though it is much less radical than the original proposal on worker participation blocked by the UK 10 years ago [7.9].

HEALTH AND SAFETY AT WORK

Health and safety is another key aspect of the social side of the internal market. Measures in this area are likely to have a significant impact on construction, which, with only 7 per cent of the Community workforce, accounts for approximately 30 per cent of fatal occupational accidents. In this area, one of the most significant pieces of legislation is likely to be the temporary and mobile sites directive, currently at proposal stage, which is specifically aimed at the construction industry.

The Single European Act introduced a new article, 118a concerned with health and safety at work, into the treaty and measures under it can be adopted on a qualified majority vote:

'*Article 118A*

(1) Member States shall pay particular attention to encouraging improvements, especially in the working environment, as regards the health and safety of workers, and shall set as their objective the harmonisation of conditions in this area while maintaining the improvements made.

(2) In order to help achieve the objective laid down in the first paragraph, the Council, acting by a qualified majority on a proposal from the Commission . . . shall adopt, by means of directives, minimum requirements for gradual implementation, having regard to the conditions and technical rules obtaining in each of the Member States.

Such directives shall avoid imposing administrative, financial and legal constraints in a way which would hold back the creation and development of small and medium-sized undertakings.'

There has been good progress in introducing directives in this area, but since the approach is close to that already applying in the UK the EC dimension will not be particularly evident to national practitioners. The UK Health and Safety Executive have stated that in implementing EC directives they intend to avoid disturbing the existing legislative framework, and to take advantage of the opportunities to reform and rationalise pre-1974 legislation.

The development of a common approach to safety and health throughout the EC, with standards in the less developed parts of the Community being brought to a common standard, will make it more straightforward to carry out cross-border work and make it easier for both construction managers and operatives to adapt to working in other Member States, provided enforcement practice is also reasonably uniform.

In 1989 the directive on the introduction of measures to encourage improvements in the safety and health of workers at work was adopted (89/391/EEC) [7.10] to be implemented by the end of 1992. This is the 'framework directive', so called because it establishes the framework of responsibilities for health and safety, much as the Health and Safety at Work etc. Act 1974 does in the UK.

The directive sets out the principles which are common to the specific safety directives or proposals, and which apply to all industries. These are that risks should be avoided; if they cannot be avoided they should be evaluated and combated at source. The dangerous should be replaced by the less dangerous. Measures which give protection to groups of workers should be given priority over measures which only protect individuals. In some respects the framework directive makes explicit the implicit requirements of existing UK safety law, for example in the requirement to carry out an assessment of risks. More recent regulations, specifically the Control of Substances Hazardous to Health (COSHH) 1988, already have the explicit requirement to carry out an assessment of risk.

Another key feature is the requirement to adapt work to the individual. The directive also explicitly refers to first-aid, fire precautions and emergency arrangements. In the UK such aspects have been covered in sector specific regulation such as the Construction Regulations. Many other provisions, such as the requirement to consult workers' representatives on health and safety measures, to provide information and training, and for employees to take responsibility for their own and others' safety and to co-operate with employers over safety matters closely parallel the existing provisions of Health and Safety at Work etc. Act 1974.

Specific health and safety requirements are contained in 'daughter' (a literal translation of a French term) directives much as regulations subsidiary to the Health and Safety at Work etc. Act are used in the UK, and are in practice being used to implement the directives. For example, the Noise at Work Regulations which came into effect on 1 January 1990 implement the Council directive 86/188/EEC (OJ L137 24 May 1986 p. 28) 'on the protection of workers from the risks related to noise at work'. This directive illustrates one difficulty caused by national differences. It requires that 'where it is not reasonably practicable to reduce the daily personal noise exposure of a worker to below 85 db(A) the worker exposed shall be able to have his hearing checked by a doctor or on the responsibility of the doctor and, if judged necessary by the doctor, by a specialist'. The regulations do not contain anything equivalent to this, because, it is argued, the tests can be carried out on the National Health Service. The unions argue that reliance on the National Health Service to carry out the checks does not meet the intentions of the directive and that employers should be obliged to provide hearing tests. The unions also claim that the emphasis on controlling noise is less in the regulations than in the directive.

There are a series of 'daughter' directives at various stages of development, several of which will have a significant effect on the construction industry.

The workplace directive (89/654/EEC)

This applies to most fixed, permanent workplaces and excludes construction sites, amongst others. However, since many of the requirements relate to buildings and other fixed structures, designers will need to be aware of the directive. There may well be implications for the conversion of existing structures. These requirements, contained in the annexes, cover structural stability, electrical installations, emergency escape routes, fire precautions, ventilation, temperature, lighting, doors and gates, dangerous areas, rest rooms and areas, changing rooms and lockers, washing facilities, toilets and first-aid equipment. New workplaces have also to meet requirements regarding windows, skylights, escalators and travelators, and loading bays. The directive was adopted on 30 November 1989 and must be implemented by the end of 1992.

Manual handling of loads directive (90/269/EEC)

This follows the standard pattern of requiring risk assessment and measures to reduce or eliminate it. Employers are required to take steps to avoid the need for manual handling, if possible, or if not, to organise the workplace so as to make manual handling as safe and healthy as possible. The directive was adopted on 29 May 1990 and must be implemented by the end of 1992.

The use of work equipment directive (89/655/EEC)

This directive imposes duties on employers to make sure that work equipment is safe and properly maintained; that, if the use of equipment involves risk, only designated persons have access to it, and that where equipment cannot be made completely safe that measures should be taken to reduce the risks. There are also provisions relating to the characteristics of the equipment, how it can be maintained, what might happen if it fails, and the emissions of gas, dust, liquid and other substances, and the spread of fire. The directive was adopted on 30 November 1989 and again implementation must be by the end of 1992.

Personal protective equipment directive (89/656/EEC)

This directive, adopted in the same meeting as the above directive, and also to be implemented by the end of 1992, has provisions on the assessment of risks, and the selection and maintenance of personal protective equipment. Workers are to be involved in its selection, and information, instruction and training in the use of the equipment are to be provided. The equipment must be provided free of charge. The annexes to the directive give specific details about the kinds of work for which personal equipment is needed and about the equipment itself.

Display screen equipment (visual display units) directive (90/270/EEC OJ L157 21 June 1990 p.14)

This directive, to be implemented by the end of 1992, has provisions on the design of equipment and its use. Periodic breaks and changes of activity are to be provided. New equipment must meet certain standards. Existing, non-complying workstations can be used for another four years. Given the rate of obsolescence for computing equipment, replacement is more likely than modification. There are also entitlements to eye tests and special visual corrective appliances (presumably spectacles), where needed.

Other directives and proposals

There are other directives or proposed directives on carcinogens, biological agents, underground mines, and the surface extraction of minerals or other materials. The latter will clearly affect companies involved in open-cast mining and quarrying. It will adapt the provisions of the Workplace Directive to the specific characteristics of the industries, covering the use of explosives for example. It is likely to be adopted towards the end of 1991.

Proposal for a 'Directive on the Implementation of Minimum Safety and Health Requirements at Temporary or Mobile Work Sites'

The proposed 'Directive on the Implementation of Minimum Safety and Health Requirements at Temporary or Mobile work Sites' [7.11] is aimed specifically at the construction industry and covers any site where construction, equipping, alteration, renovation, repair, upkeep, maintenance and demolition of any type of building or structure takes place. It breaks new ground by requiring the appointment by the client of a project supervisor responsible during the design stage for ensuring that safety and health are planned into a project right from the inception stage to completion. This function could be carried out by a developer, or the engineer or architect or any other appropriate person. A project supervisor must also be appointed for the execution stage, to have responsibility for the co-ordination, supervision and inspection of work from the point of view of health and safety. The directive will require plans dealing with safety and health to be produced according to the degree of risk.

According to the Federation of Civil Engineering Contractors in the UK the earlier drafts of the proposal reflected serious misunderstanding about roles and relationships in the construction industry.

The proposal includes annexes which contain detailed requirements related to specific hazards found in the industry. The requirement to appoint project supervisors would make explicit the allocation of responsibility under the Health and Safety at Work etc. Act 1974. There may be some conflict with the requirement to appoint a safety supervisor under the Construction Regulations, but these regulations, which date back to the 1960s are likely to be replaced by the regulations which implement the directive. A 'common position' on the proposal is expected in autumn 1991.

Other measures

Other health and safety initiatives described in the Action Programme relating to the implementation of the Social Charter [7.12] cover the following:

- signs about health and safety at work;
- information about dangerous industrial agents at work;
- protection of workers from the risks related to asbestos at work;
- health and safety requirements in the transport sector;
- exposure to physical agents such as vibration and electromagnetic radiation;
- the establishment of an agency in the field.

Impact of EC health and safety legislation

The emphasis on protecting the consumer and on the health and safety of workers, including those of sub-contractors and the self-employed, is likely to continue the pattern already seen over recent years in UK legislation and case law. There is, however, a possible shift of emphasis towards requiring more explicit planning and risk evaluation, together with clearer allocation of responsibility. From the point of view of the developer, designer, contractor and sub-contractor carrying out work in other Member States, there is likely to be a development of a common culture of safety which will make such work easier to adapt to. In addition, the introduction of common European requirements for safety precautions and common standards for protective equipment will also make cross-border work less of a problem.

National variations in health and safety enforcement [7.13]

A measure of uniformity is being introduced in health and safety rules and approaches to responsibility, but the same cannot be said about enforcement. Only Denmark and the UK have a single government body responsible for enforcement of largely coherent bodies of legislation.

In France there are many agencies responsible for this area; roles often overlap and the agencies may have other functions as well. The Services Exterieures du Travail et de l'Emploi also has responsibilities for industrial relations, rather like that of the UK's Arbitration and Conciliation Service. It is only concerned with the employers and the employed, not with the self-employed.

In Germany each of the Länder (11 in western Germany and five in the east) has its own health and safety inspectorate or *Gewerbeaufsicht* (GAs), whose duties extend well beyond health and safety into conditions of employment and even environmental matters. In addition, there is a network (95 in western Germany) of industry co-operative and insurance bodies, or *Berufsgenossenshaften* (BGs), operating in specific industries. These are funded by employers' contributions but have the power to make legally binding regulations for accident prevention. According to Teague and Brewster [7.14]

> 'No other Member State gives such a prominent role to insurance in the setting and enforcement of occupational health and safety standards. One direct consequence of the regulatory powers of the BGs is that the total volume of health and safety legislation is much greater than in Britain and much more detailed. Both GA and BG inspectors spend much of their time on visits . . . both have powers to issue administrative fines for minor offences, to issue notices and ultimately to ask the public prosecutor's office to pursue offenders through the courts.'

Chapter 8

Environmental Policy

The Community has been particularly active in environmental policy. The pace is to some extent set by Germany and the other northern Member States. Measures have affected other industries more than construction but the problems, potential problems and Community policy in response are likely to have a major impact on the construction industry, providing opportunities for consultants and contractors who have prepared themselves.

The environment and the construction industry

The construction industry has tended to be sceptical about environmental concerns; environmentalists have often been obstructive to construction projects, for example because a rare species of plant or bird was threatened. Environmentalists have been strongly opposed to road construction and housing developments which extend towns into the countryside. The construction industry's traditional antagonism towards the environmental movement is now softening as environmental problems are becoming worrying. The impact of buildings, transport systems, hydroelectric projects, land reclamation, tidal power schemes, water supply and waste water treatment on the environment is clear. Construction products are also significant because of their possible effect on health and because they consume raw materials on a large scale. Approximately 50 per cent of carbon dioxide emissions, the most significant greenhouse gas, is associated with dwellings, primarily from water and space heating. A further 20 per cent of carbon dioxide emission is due to transport, with implications for transport planning and road construction. Most important of all perhaps, measures connected with concern for the environment could be a major source of work for the construction industry.

At a series of discussion meetings in the first half of 1990 held by the UK Construction Industry Research and Information Association

(CIRIA) on the construction industry and the environment, the issues were seen as global, regional and local. Global issues included the greenhouse effect, the damage to the ozone layer, depletion of resources and large-scale pollution, particularly through discharge of waste containing chemicals. Regional issues identified were changing patterns of use, particularly affecting transport, energy use, town planning, heating and cooling of buildings, use of raw materials in short supply or involving the production of chemical waste (use of chromium in building products was mentioned) and the re-use and recycling of building materials. Local issues related to the effect of buildings on their inhabitants – movement towards the medically benign building, with the identification of past problems including asbestos, 'sick building syndrome', Legionnaires' disease and radon accumulation in buildings.

The meetings discussed the failure of the industry to respond appropriately to issues at an early stage, and referred to the Montreal Convention (1987) in which drastic cuts in the use of CFCs were agreed, creating problems for an unprepared industry.

Impacts

The developments and changes resulting from measures to protect the environment and, in the longer term, the potential impacts of environmental changes are enormous, possibly even cataclysmic. These will have considerable implications for the type and amount of work to be done by the construction industry. Companies may react defensively to measures which could reduce traditional sources of work such as road construction, but other opportunities are being created. Some companies are responding, not always honestly, by promoting themselves and their products as 'green'. Others are seeing new markets and are conducting the research and development necessary to exploit them. The CIRIA meeting called on the industry to establish a means to identify issues as early as possible, to co-ordinate responses, to carry out research and testing and to disseminate the results to the public, clients, government and the industry itself. Not only were reactive responses required, for example commenting on and criticising proposed government legislation and preparing to meet demands, but also pro-active responses, actively seeking solutions to emerging problems and proposing solutions.

Economic growth, trade and the environment

The Community single market is expected to accelerate growth with an impact on the environment, particularly though consumption of energy and raw materials. The relationship between growth and energy is not direct and there are wide variations between countries. For example, demand for fossil fuels per head in Japan is 40 per cent of that in USA while GDP per head is nearly 20 per cent greater. Use of fossil fuels in the Soviet Union is 71 per cent that of the USA level, despite a much smaller output per head. GDP in the UK increased between 1979 and 1989 by 25 per cent without an increase in energy use.

Growth, however, tends to result in increased use of energy, which can mean increased production of carbon dioxide or nuclear waste material and more acid rain. The generation of solid and liquid waste, increased demand for transport, particularly by car, destruction of the countryside and natural ecosystems as land is taken over and developed are other problems.

As environmental concern grows in the countries which feel they can afford to care there also is a fear that tighter regulations, or the imposition of taxes designed to deter pollution, will result in the export of jobs (along with the pollution) to countries with lower standards – so-called 'environmental dumping'. To prevent this from happening, the environmentally concerned Member States are determined that the same standards of environmental protection, with their consequences for production costs, should apply throughout the Community, and beyond if possible. The concern about high environmental standards causing job losses is partially offset since the development of clean technologies to meet the high standards will offer export potential as environmental concern spreads to other countries, or they are otherwise coerced into following suit.

The environment and international trade

The Community can harmonise environmental standards related to industrial production within the Member States, but the problem of environmental dumping in international trade remains. The GATT agreement contains no environmental provisions at present, and even the draft texts being discussed within the context of the Uruguay round do not deal with the problem. In the longer term, such agreements are likely to allow countries to act against imports produced by environmentally damaging means or in countries without appropriate regulation, for example without limits on greenhouse gas emissions. The political problems could be immense, and such policies may have to be linked with aid to less developed countries to assist in the introduction of cleaner technologies.

Views on the environment amongst Member States

In the UK it is hard to appreciate the level of concern over the environment on the Continent. To the British public, green issues are often associated with birds, hedges, ponds and lead in petrol. Environmental damage is not associated with harm to health or income. Southwesterly winds, coming off a broad expanse of Atlantic Ocean, protect the British Isles against airborne pollution. The ocean also provides a buffer against marine pollution and even the climate is largely protected from extremes of heat or cold. British rivers do not cross frontiers and are short and fast in comparison with the major continental rivers, with less prospect of accumulating pollution. The British environment appears to be in robust health. It apparently recovered rapidly and completely after the serious levels of river and air pollution in towns in the first half of the century. Smogs had been a serious problem, with the worst of these resulting in 4000 deaths and widespread breathing problems in December 1952. The Clean Air Acts of 1957 and 1968 controlled the problem, though urban air pollution is again becoming a concern with clear links to illness and early death [8.1].

Green issues are much higher on the agenda on the Continent, particularly in Germany, Denmark and Holland. The success of the Green party in Germany is well known (8 per cent of the vote in 1987) and is not merely the result of its electoral system allocating a proportion of parliamentary seats according to share of the vote. Since the early 1980s there has been intense concern over dying forests. Germany is more heavily forested than the UK and the forests are more widely spread around the country, so that damage is obvious to all. In the UK the majority may never have seen a real forest. In Germany, in the Ruhr and Berlin, smog warnings have been issued in the last decade and only essential users were allowed to use their cars.

Many German rivers are heavily polluted. The Elbe, which runs through what used to be East Germany, used to carry dangerous levels of heavy metal pollutants as well as high levels of fertilizer and pesticide residues. After one industrial accident in Switzerland the Rhine ran red and most life in the river was wiped out. At the International Panel on Climate Change in Geneva in November 1990 it was Germany which had the most ambitious target for reduction of greenhouse gases, 25 per cent from 1987 levels by 2005 for west Germany and 30 per cent for the whole of the country (the difference accounted for by the low levels of energy efficiency in the east at unification). The UK, the fifth largest producer, meanwhile planned to achieve stabilisation at 1990 levels by 2005. Neither the USA or the USSR, which together account for over 40 per cent of human-produced carbon dioxide, were in favour of emission controls. An ironic consequence of membership of the Community is that both Holland and Germany have had problems with Community law in offering tax incentives to customers purchasing vehicles meeting higher environmental standards than required by EC law. The Dutch case involved tax inducements on cars meeting USA emission standards, while the Germans plan special incentives for diesel engined cars emitting 0.08 grams of soot per cubic metre in the face of a recently agreed Community limit of 0.14 grams per cubic metre. The German case is confused by arguments as to whether the lower limit favours German manufacturers since other manufacturers may be less advanced in the development of engines which meet the tougher limit.

Holland has one of the most ambitious programmes on the environment. Acute soil acidification has occurred in the country and there is widespread concern about pollution in the Rhine and in the North Sea. Any rise in sea levels resulting from global warming would be both particularly costly and dangerous for the country.

In northern Italy river pollution is a problem. Algae growth in the Adriatic, probably resulting from pollution of the rivers feeding into it, turned the sea red and slimy in 1990, keeping large numbers of tourists away. In Greece extreme heat and air pollution have combined to cause many deaths, mainly in Athens where much of the Greek population is concentrated. Ancient monuments in Greece have suffered badly from air pollution.

The Spanish tourist industry has been hit, partly by concern about marine pollution caused by disposing of increasing amounts of sewage into the Mediterranean. In Portugal the environment was a major issue in the elections for the European Parliament in 1989. Farmers demonstrated about eucalyptus plantations, claiming that they impoverish the soil, use too much water and that the planting of eucalyptus plantations is breaking up communities based on traditional farming. There was also concern that pollution controls in the northern Member States of the EC were driving the worst polluters in their direction. From minimal legislation the government announced a programme of measures as advanced as any in the EC, for example dealing with waste disposal and noise. The regulations go beyond EC requirements and industrial firms setting up in Portugal have to demonstrate in advance that they will be able to comply.

Pollution and green issues are perceived as crucial in most Member States. The cross-border nature of pollution and the link with trade means that the Community, and the Council of Europe (the 22 member body based in Strasbourg, not to be confused with the Council of Ministers of the European Communities) are seen as appropriate bodies at which to agree policy. Global agreements are also needed, such the Montreal Protocol on chlorofluorocarbons (CFCs) which the European Community signed in September 1987 along with 21 nations.

Community record on the environment

The Community has a record of concern for the environment with about 280 measures adopted between 1973 and 1990. These include regulations and directives in the areas of:

- Water quality, covering such aspects as drinking water, bathing water and groundwater.
- Waste, covering organisation of waste disposal within the Member States including the requirements for undertakings which treat, store or tip waste commercially to obtain permits, and the encouragement for prevention, recycling and processing of waste and the extraction of raw materials and energy from waste.
- Air, with measures relating to reducing the emissions from power stations and other large combustion plants, emissions from industrial plants and vehicles and monitoring of forest damage.
- Chemicals, with measures concerned with major accident hazards, chlorofluorocarbons (CFCs), worker protection, paints and solvents and asbestos.
- Wildlife and countryside, with measures covering environmentally sensitive areas, birds and their habitats, whales, seals, trade in endangered species and countryside protection in agriculturally less favoured areas.
- Noise, with measures covering vehicles, construction plant and noise in the workplace.
- Assessment, information and finance dealing with environmental impact assessment statements (see p 140), freedom of access to information on the environment, an experimental project on gathering and co-ordinating information on the state of the

environment, and finance to be made available for demonstration projects to develop clean technologies amongst others.

The impact of environmental measures developed at Community level and those agreed in international conventions, at which the Community negotiates on behalf of the Member States, is likely to accelerate, with the pace set by the northern mainland countries, particularly Germany and Holland.

The environment and the Single European Act

The Single European Act introduced three new Articles, 130r, 130s, and 130t in a new 'title' labelled 'environment'. Community objectives in the area are spelled out in Article 130r as being 'to preserve, protect and improve the quality of the environment, to contribute towards protecting human health and to ensure a prudent and rational utilisation of resources'. The Articles oblige the Community to take action when action at Community level is appropriate, but without limiting the freedom of Member States to introduce other 'protective' measures, which may be more stringent. Unanimity in the Council on measures introduced under these articles is required, but in many cases Article 100a, which is connected with measures in connection with the establishment and functioning of the internal market, may be the basis of action, in which case qualified majority voting is allowed. Article 100a states that proposals concerning environmental protection, amongst others, 'shall take as a base a high level of protection'.

Surprisingly, the White Paper on completing the internal market contains no indications of new directions in environmental policy, though established areas such as consumer protection and vehicle emissions, other than carbon dioxide emissions, were mentioned.

Recent developments

Community policy is based on the principles that preventative action should be taken wherever possible, that priority should be given to rectifying environmental damage at source, that the polluter should pay and that environmental protection requirements should be built into all proposals. Policy and proposals for forthcoming measures are generally spelled out in 'action plans' covering a specific period. The fourth action programme on the environment covered the period 1987–92 and a fifth action programme is in preparation. While the early action programmes tended to deal with specific problem areas, the later programmes have shifted the emphasis to prevention – the recent directive on environmental impact assessments (see below) being an example of a measure designed to prevent problems from arising.

Environmental impact assessments

Of particular interest to the construction industry is directive 85/337/EEC [8.2] which required Member States to introduce mandatory assessment of the environmental impacts (EIAs) of various public and private projects at the planning stage. There are two categories of project referred to. The first for which an environmental impact assessment must be made unless covered by a specific exemption covers oil refineries, power stations, radioactive waste storage and disposal facilities, iron and steel works, asbestos works, chemical installations, motorways, express roads, railways and airports, trading ports and inland waterways and installations for incineration, treatment or landfill of hazardous waste. The second category requires assessment if there is likely to be a significant impact on the environment and includes a variety of industrial facilities, infrastructure and other projects. In the UK the directive was implemented by a series of regulations in 1988. The environmental impact assessment is required to deal with the impacts on human beings, fauna and flora, soil, water, climate, landscape, material assets and cultural heritage. The environmental impact assessments are to be provided for use of the authorities, but the public will also have access and must be consulted.

The requirement demonstrates a faith in the preventative effect of requiring information to be made available to the planning authorities, project promoters and the public. There is no specific requirement to take action on environmental impact statements, but they will be considered alongside other information by the planning authorities in reaching decisions on whether to approve a scheme. A variety of consultants have recruited and adapted existing expertise to offer the facilities for undertaking environmental impact statements. The assessments have generally been helpful, with early consideration of the environment causing developers and others to modify schemes, often at little expense. Protesters are provided with scientific assessment data, which are not always helpful to their cause. The disadvantage is the added expense and delay, particularly where the environmental impacts are complex.

Freedom of access to information about the environment

Another measure which will increase public involvement in the projects and processes with environmental impact is Directive 90/313/EEC on the freedom of access to information on the environment [8.3]. Under this measure, which is to be implemented by 31 December 1992, bodies with public responsibilities for the environment and under the control of public authorities are to provide information about the environment on request unless the 'confidentiality of the proceedings of public authorities, international relations, national defence, public security, etc.' are affected. Any refusal should carry the right to obtain judicial or administrative review. Another requirement is that Member States should publish periodic reviews on the state of the environment.

European Environmental Agency

A significant development is the creation of a European Environmental Agency under Council Regulation 90/1210/EEC [8.4]. The primary role of the agency will be to provide the Commission with the necessary scientific and technical information to guide environmental policy and legislation. It is also charged with promoting the development of methods for forecasting environmental impacts, assessing the cost of damage to the environment and the costs of preventative action. It will not be given a policing role, though its monitoring measurements could prove useful to national enforcement agencies or the Commission. The agency is also likely to have to co-ordinate with east European countries where so much damage has been done. It is not yet decided where the agency is to be based, but Cambridge has been mentioned as a possibility.

Strict liability for damage caused by waste

Another measure is designed to increase consumer protection and to reinforce the polluter-should-pay principle. This proposal [8.5] seeks to impose strict liability on waste producers and others responsible for waste. The implications are discussed in chapter 10.

Prospects

Existing, proposed and likely measures connected with the environment will have a significant impact on demand for construction and the type of work to be done. The water, waste, electricity and transport sectors are all affected.

The water industry throughout the Community is having to embark on a substantial programme of investment even to meet existing standards. When the UK water industry was privatised in 1990, drinking water in many parts of the country failed to meet Community standards, for example on concentrations of nitrates. A fifth of the sewage treatment works were deficient and nearly half the designated bathing waters failed to meet the standards. The Department of the Environment estimated that £12 billion would have to be spent by the water industry in England and Wales. The work involved includes taking measures to reduce pollution of aquifers, adding treatment facilities to sea outfalls and making alternative arrangements to replace sewage sludge dumping at sea, which must be phased out by 1998.

A proposal for a directive on municipal waste water treatment [8.6] issued in December 1989 will require secondary treatment of waste water as the norm, and discharges into 'sensitive' water courses, such as those already contaminated by fertilisers and agricultural wastes, will require tertiary treatment. The Department of the Environment estimates that requirements relating to sensitive coastal and estuarial locations would cost £1.5 billion to install and £45 million per annum to run. Scotland has a particular problem, with many small communities discharging sewage into the sea following only primary treatment. Many councils will be forced to seek help, some of which will be available under the various grants and loan schemes organised by the Community. Spain and Portugal have similar problems, exacerbated by growing indigenous populations supplemented by visitors and retired people. Another effect of rising standards will be higher charges for industry in treating and disposing of waste water.

Transport is subject to a number of environmental influences. Throughout the Community there is concern about congestion, environmental impact of roads on towns and in the country, the levels of airborne pollution in towns particularly, and also in the contribution to acid rain. Concern about greenhouse gases is another factor. Higher exhaust emission standards being introduced for cars, buses and lorries will alleviate the health hazards and the problem of acid rain, but with unchanged transportation policies the road congestion problem seems likely to become even more intolerable. Already there has been a resurgence of interest in light rail systems and other forms of public transport. To be effective in restoring the quality of the urban environment these schemes must be combined with pedestrianised areas and other limits on car use in towns. Once cars become less useful within towns there may be further shift of demand from inter-urban roads towards rail. Even the shape of towns may be influenced, with a return to compact towns centred around stations, and concentrated on bus and tram routes. Studies are already being conducted on the relationship between choice of housing and employment location and transport mode. A shift to rail transport for freight is also likely, particularly with the development of a Europe-wide network with links to the UK and Spain. Industry could move back to a preference for sites adjacent to railway lines, rather than the modern tendency to select sites near motorway routes. Some return to inland waterway freight transport is also possible.

Industries dealing with waste will be affected. The waste disposal industry is already substantial with western Europe spending 24 billion ECU in 1989 for the disposal of solid waste, 60 per cent of it industrial refuse. The priority will be capital investment in new treatment and recycling facilities of all types, including those used to generate power from waste. The recycling of building waste is one of the priorities. The treatment and restoration of land will be another priority. At present approximately 50 per cent of all new development in the UK takes place on old sites and this is likely to increase. The 1988 Derelict Land Survey in the UK showed that about 40 000 hectares of land was derelict and that about three quarters of this justified reclamation measures [8.7]. Landfill areas particularly present problems such as dealing with contamination to water supplies and the generation of gases, predominantly methane (a powerful greenhouse gas). There is enormous scope for applying expertise in this area in eastern Germany and in central and eastern European countries.

The greenhouse effect

Global warming is a fact, but the particular influence of human produced (anthropogenic) greenhouse gases, which include carbon dioxide, methane, nitrous oxide and the chlorofluorocarbons (CFCs), is not yet certain. There has been a mean global temperature increase of 0.4°C over the last century. Atmospheric models suggest a figure closer to 0.8°C over the same period. The difference could be due to imprecision in the models, or some natural variation, such as in solar radiation, which could be masking the true influence of the anthropogenic greenhouse gases. A report produced by around 200 climatologists for the Intergovernmental Panel on Climate Change (IGCC) in November 1990 predicted that if present trends continue

'– global mean temperature would increase by 0.3°C per decade (within the range 0.2°C to 0.5°C per decade); this is greater than the world has seen over the past 10 000 years. Global mean temperatures over the next century would become higher than at any time over the last 150 000 years; sea-level would rise by 6 cm per decade (within the range 3–10 cm per decade) over the next century, three to six times the rate seen over the last 100 years;
– rapid changes in climate will change the composition of natural living systems; some species will benefit, while others will not be able to migrate or adapt fast enough and may become extinct; and
– the effect of warming on biological and chemical processes seems likely to increase the atmospheric concentrations of natural greenhouse gases in a warmer world.' [8.8]

Rises in sea level, primarily from thermal expansion of the oceans and melting of land ice, could submerge low-lying land and increase ground water salinity. Storms and weather extremes could become more frequent and intense. Vegetation and natural ecosystems could fail to adapt rapidly enough and some land could become infertile. Diseases could spread and become a severe problem, particularly insect borne diseases, since insects are likely to benefit from climate change. Some of these changes could result in famine and mass human migration.

Most developed countries accept the need to take action to counter the potentially catastrophic changes which could come about. Action will have to be taken before the scientific evidence is irrefutable, because the delays and lags in the systems are so great that delayed action would have to be much more drastic and the consequences could be even more severe. Unfortunately the governments of the USA and the USSR, the major producers of carbon dioxide, would both face political problems in introducing measures to restrict the carbon dioxide production and are not yet willing to do so. President Bush has succeeded in finding scientists to disagree with the findings of the IGCC panels, but other scientists give little credence to their work [8.9]. In any case, the burden of reducing output should fall on the developed countries who produce more per head. The less developed countries face the additional problem of rapidly increasing populations, which impose pressures of every kind on the environment. The Commission and the governments of the Community are already committed to, or have proposed, curbs on greenhouse gas production which will have considerable impacts on the construction industry.

The areas subject to most impact are coastal defence works, energy generation by renewable technologies, transport, building materials, energy conservation and design standards. In addition, the impact of measures to reduce the production of greenhouse gases could result in reduced economic growth, though some argue that the changes required could generate some forms of growth. If adaptation is too slow and the forecasts are proved even partially correct, drastic action could be called for, possibly in the face of widespread economic dislocation.

Action on the greenhouse effect would include both fiscal and regulatory measures. The most favoured fiscal measure is a 'carbon tax' on fossil fuels. Mr Carlo Ripa di Meana, the environment commissioner, is known to favour a carbon tax, and at one stage proposed that the level should be set by the Commission, with the proceeds being used to pay for the preservation of tropical rain forests. The Organisation for Co-operation and Development (OECD) has also emphasised the role of taxes in the fight against pollution. The Paris-based International Energy Agency has urged the use of onerous taxes on carbon fuels, and has spoken of the need to double the price of coal, and increase energy prices by at least 20 per cent.

Regulations could require the energy efficiency of buildings and their fittings to be sharply increased; there is already research, particularly in North America, Scandinavia, Holland and Germany, on construction details for 'superinsulated' houses which can reduce the annual space heating demand from 56 gigajoules (GJ) for the typical new UK house (1990 building regulations) to 6 GJ. In Germany, the UK and elsewhere, research has developed transparent insulation materials (TIMs) which insulate and trap solar energy. Kaiser Bautechnik and Okalux in Germany will soon start mass producing these materials. Rating and publication of the energy efficiency and environmental 'friendliness' of buildings and of the sustainability of production of components may be required. The use of tropical hardwoods has been criticised because much production is not sustainable, but wood as a building material has much to offer. Little research is available on such issues but the impact on design and construction could be considerable. In some cases environmentally motivated actions may have been counter-productive.

The economy will be transformed, with low energy goods and services being relatively favoured. Building costs will increase, reducing demand, but existing buildings will need to be replaced or converted to achieve higher insulation. Steel and concrete are energy intensive materials and will give ground to timber products. The use of timber is an effective way of taking carbon out of the carbon cycle, it requires little energy for harvesting and processing and forests are environmentally desirable. The raw materials for the production of steel and concrete require mining or quarrying.

Building design and town planning will need to change, particularly with emphasis shifting from planning for increasing use and ownership of cars. Growth in public transport would favour the Victorian, or even medieval town planning – compact towns ('jewels on a necklace'), with fingers of development along public transport routes and green spaces in-between, rather than the suburban sprawl containing detached houses each with garden and garage which has become the norm. Apartment blocks might come back into fashion in the UK.

Renewed enthusiasm will be seen for combined heat and power schemes, creating a link between urban areas and power station, as well as for the renewable energy sources – wind, wave, solar, tidal and geothermal. Whether nuclear energy will be boosted remains debatable.

If the more gloomy scenarios are realised there could also be a sharply increased workload in providing large scale sea-defence works, increased water storage facilities to cope with drier, warmer summers and greater variability of weather, and in some regions housing to cope with migrations from areas rendered inhospitable by climate change. This would probably take place in the context of economies struggling to cope with rapid change and agricultural disruption, with disturbing political implications.

Chapter 9

Standards and Codes

Standardisation is a key aspect of the single market. Europeanisation of standards is of particular interest to companies wishing to market goods and services in more than one European country. The problem of differing standards is rated by business as one of the most important barriers to be removed in the creation of the single market. One of the sections of the 'cost of non-Europe' study [9.1] described in chapter 1 examined the construction products sector. The authors estimated that the Community market for construction products was worth about ECU 110 billion per annum, and that the cost of market fragmentation in the 12 Member States was about ECU 820 million per annum. For the five largest Member States the study estimated that long-term benefits, which could come about as the industries improved their economies of scale, could be as much as ECU 1.7 billion. The study found widely varying degrees of import penetration in the five largest Member States: 15 per cent in Italy, 20 per cent in Germany, 30 per cent in France and 50 per cent in the UK. It commented that the demanding standards imposed in the northern Member States, particularly France and Germany, made exporting into those countries particularly difficult. This chapter examines the developments in the way the Community has tackled the problem of standardisation, the European standardisation bodies, and general progress before moving onto the Construction Products Directive. This Directive was due for implementation in the Member States in July 1991, and should have important consequences for suppliers, but progress towards harmonised (i.e. European) standards is likely to be very slow. The design codes, called Eurocodes, which will eventually replace national design codes are also considered, as are developments in the fields of testing and certification and quality assurance. The chapter also makes brief reference to a recent policy document issued by the Commission in the area of standardisation.

Non-tariff barriers and quotas

Initially, tariff barriers and quantitative restrictions, or quotas, were thought to be the main factors keeping the European market fragmented. Work was done in that direction with much success but it rapidly became evident that the problem was broader. It was clear that standards can, either intentionally or unintentionally, isolate national markets from full competition. For instance, differing standards can oblige manufacturers wanting to export to introduce many product variations and to obtain a number of approvals. The cost, time and administrative effort involved deters and may harm competition. This compartmentalisation of the European market denies opportunities for economic efficiency and technological development and reduces export possibilities. Since standards apply not only to the products, but also to their testing, mode of inspection, certification and methods of production, the problem is exacerbated.

The prescriptive approach

Technical harmonisation has been a goal from the inception of the Community. The original approach was to try to hammer out all variation and aim for a beautiful set of detailed product specifications applicable throughout the Community. The difficulties of obtaining agreement at that level of detail led to the eventual abandonment of this approach.

The Cassis de Dijon case

A case in 1979 helped develop a new approach to harmonisation – the *Cassis de Dijon* case [9.2]. A German company had found that it was prevented by German law from importing *Cassis de Dijon*, a popular French blackcurrant liqueur, because it infringed German standards on minimum alcohol content. The Court of Justice held that the German government could only prevent the sale of the drink if it could demonstrate that the liqueur was harmful to health or if it did not comply with consumer protection or tax laws.

The court underlined that the aim of all national technical standards and regulations is largely the same – the protection of public health and safety. The Commission subsequently gave up the original prescriptive approach to adopt one which laid emphasis on fundamental objectives. The Internal Market White Paper restated the new approach in the following terms:

> 'The general principle should be approved that, if a product is lawfully manufactured and marketed in one Member State, there is no reason why it should not move freely throughout the Community. The objectives of national legislation in this respect are essentially identical. Although they may take different forms, they amount to the same thing.'

The 'new approach' to technical harmonisation

In 1985, the 'new approach to technical harmonisation' was introduced through Council resolution 85/C136/01 [9.3]. The new approach reduces the process of harmonisation to the fundamentals referred to in the *Cassis de Dijon* case. The new approach directives specify in general terms the essential requirements for products, primarily in relation to health and safety. European standards then provide fuller detail of the requirements, but are still generally expressed in terms of performance requirements. Products which meet the essential requirements, as elaborated in the European standards, or otherwise, should carry the CE (Communauté Européenne) mark. Member States must not impede the free mobility of such products. It will be a criminal offence to sell products which do not comply with the rules (with enforcement through product liability legislation - in the UK, the Consumer Protection Act 1987). Under Article 100a of the Single European Act the Commission is to 'take as a base a high level of protection' in its proposals, including proposals for new approach directives concerned with standards, in relation to health, safety, environmental protection and consumer protection. The directives set minimum levels, but these are required to be at a high level.

The new approach directives follow a largely common pattern. The range of products covered is generally broad. The essential requirements that products must meet and the methods of specifying those requirements are described. The method by which manufacturers show that their products meet the requirements are also covered. This is normally by means of an 'attestation' which can include a declaration by the manufacturer, may be backed by test results, or by a certificate or test results produced by an approved independent body. The procedures for attestation under the Construction Products Directive are described below.

New approach directives dealing with simple pressure vessels, toy safety (look out for the CE mark), electromagnetic compatibility, machinery safety, gas appliances and construction products have, amongst others, been adopted.

New measures

A significant measure relating to the attempt to increase the pace of harmonisation of standards is directive 83/189/EEC [9.4] on mutual information procedures, which came into effect in January 1985. This requires national standards organisations to provide the Commission, and CEN or CENELEC, with copies of new technical regulations in draft form and apply a three month standstill on the implementation of those drafts. The standstill period can be extended if either the Commission or a Member State consider that the technical regulation would constitute a barrier to trade or if a directive or regulation relating to the matter exists or is proposed. By 1989, 30 proposals for technical regulations in the Member States had been blocked by the Commission following this procedure. CEN and CENELEC also have internal standstill agreements with their members who agree not to introduce new standards when a CEN or CENELEC standard is proposed.

CEN and CENELEC

CEN (Comité Européen de Normalisation), the European Committee for Standardisation, is the association of the national standards bodies of the EC Member States and the EFTA countries, established in 1961. A year later a similar organisation was created which later became known as CENELEC, the European Committee for Electrotechnical Standardisation.

CEN and CENELEC deal with all questions relating to standardisation activity at the European level with their budgets, internal rules, work programmes and allocation of resources being agreed by the national standards organisations under a voting procedure. In contrast to the national organisations, only professional interests are directly represented on the governing bodies of CEN and CENELEC. Those such as public authorities, manufacturers, or other users of standards are not directly represented.

Since 1985 a majority voting procedure was introduced in CEN and CENELEC for new standards. Also, with the Single European Act of 1987, a qualified majority voting procedure, on the same lines as that used in the Council of Ministers, was introduced for standards which the Commission has instructed CEN or CENELEC to produce under a mandate. If such a mandated standard is agreed in CEN or CENELEC, even if not unanimously, it must be adopted by all the Member States of the Community, and the equivalent national standard, if any, withdrawn. If the standard does not receive a favourable vote following voting by the full CEN or CENELEC membership, the votes of members from the Community Member States are counted separately. A qualified majority amongst the Member States for adoption of the standard then requires all Member States, and the EFTA countries which voted in favour, to implement the standard. The relationship between CEN and CENELEC and the European Community is established in a memorandum of association [9.5].

Progress

The rate of adoption of standards in CEN and CENELEC has increased. In 1989 CEN adopted 130 standards as against 19 in 1982, while CENELEC adopted 126 in 1989 as against 20 in 1982. Currently about 1250 European standardisation documents have been adopted. There are 20 000 standards in Germany, 13 000 in France and 10 000 in the UK. Much work remains to be done, particularly in the field of construction where there are few common standards.

The Construction Products Directive

The Construction Products Directive 89/106/EEC [9.6], adopted on 21 December 1988 and due to have been implemented by 27 June 1991, is likely to have a significant impact on cross-border trade and on national practice. The comprehensive scope of the directive should avoid the need for another directive in this field. It does not impose detailed technical requirements but provides procedures designed to allow harmonisation to be introduced by stages. Six essential requirements concerned with the safety of people and property are defined together with a mechanism for elaborating these requirements, which will be by means of 'interpretative documents'. The directive also sets out the procedures to be used for developing harmonised standards and technical approvals, and for demonstrating compliance with the essential requirements. Construction products which meet the six essential requirements will carry the CE (Communauté Européenne) mark.

Coverage

The definition of a construction product is wide-ranging. A construction product is regarded as 'any product which is produced for incorporation in a permanent manner in construction works, including both building and civil engineering works, in so far as the essential requirements relate to them'. It embraces any items which participate in the functioning of a building (including mechanical and electrical work). Products such as scaffolding and formwork appear to be outside the scope of the Construction Products Directive but should be covered by the directive (at the proposal stage at the time of writing) on 'implementation of minimum safety and health requirements at temporary and mobile work sites' (see chapter 7).

Essential requirements

The Construction Products Directive requires the Member States to make sure that construction products are 'placed on the market' only if they are fit for their purpose. To be fit in the context of the Construction Products Directive means to have 'such characteristics that the works in which they are incorporated assembled, applied, or installed, can, if properly designed and built, satisfy the essential requirements'. This is an unusual feature of the directive – the essential requirements relate not to the subject of the directive, construction products, themselves, but to the structures into which they are incorporated. The scope of the directive thus extends to designers and contractors as well as product manufacturers.

The six essential requirements describe what is imperative for health and safety but do not explain how these are to be achieved. There are no rules covering production, assembly or installation. The requirements are expressed in very general terms, although some elaboration is given in the annex.

The six essential requirements are (largely in the words of the directive):

(1) Mechanical resistance and stability. The construction works must be designed and built in such a way that the loadings likely to act on it during its construction and use will not lead to the total or partial collapse of the work, inadmissible deformations, damage to

other parts of the works or to installed equipments or to fittings due to the deformation of the load-bearing construction, damage by an event to a degree disproportionate to the original cause.

(2) Safety in case of fire. The construction must be designed and built in such a manner that in the event of an outbreak of fire the load-bearing capacity of the construction can be assumed for a specific elapse of time, the generation and spread of fire and smoke within the works are limited, the spread of fire to neighbouring construction works is limited, occupants can leave the works or be rescued by other means. The safety of rescue teams is to be considered.

(3) Hygiene, health and the environment. The construction work must be designed and built in such a way that it will not be a threat to the hygiene or health of the occupants or neighbours, in particular as a result of emanation of toxic gas, the presence of hazardous particles or gases in the air, the emission of dangerous radiation, pollution or poisoning of the water or soil, defective disposal of waste water, smoke, solid or liquid wastes, the presence of damp in parts of the works.

(4) Safety in use. The construction must be designed and built in such a manner that it does not present unacceptable risks of accidents in service or in operation such as slipping, falling, collision, burns, electrocution and injury from explosion.

(5) Protection against noise. The construction must be designed and built in such a way that noise perceived by occupants or people nearby is kept below a level that will not threaten their health and will allow them to sleep, rest and work in satisfactory conditions.

(6) Energy economy and heat retention. The construction works and the heating, cooling, ventilation installations must be designed and built in such a manner that the amount of energy needed in use will be low, having regard to the climatic conditions of the location and the occupants.

Products must provide for the essential requirements to be satisfied for an 'economically reasonable working life', subject to normal maintenance and foreseeable working loads (termed actions) and requirements.

Interpretative documents

A set of 'interpretative documents' are being developed under the instructions of the Commission to enlarge on and clarify the essential requirements. They will be the basis for preparing the new harmonised standards, guidelines for European technical approvals and for recognition of national technical specifications.

The interpretative documents allow for different classes of performance depending on geographical and climatic conditions, patterns of use and required levels of protection. The interpretative documents will set out the technical basis of, and criteria for, the system of classification and indicate methods of correlating these classes or levels of requirement with standards or technical approvals.

Technical specifications

Construction products are said to be fit for their intended use if they comply with the six essential requirements; in which case they can carry the CE mark. They must comply with one of the following, which are collectively referred as technical specifications:

(1) National technical standards transposing harmonised standards (i.e. European standards). They are published in the UK as identically worded British Standards.
(2) A European technical approval; this is a form of technical assessment certifying compliance with the essential requirements. This is an agrément procedure designed for products of an innovatory nature for which harmonised standards are not possible.
(3) Any other national technical specification, if no harmonised specification exists, provided it is recognised as complying with the essential requirements. These other technical specifications may include non-harmonised European or other international specification adopted at national level.

There is a procedure in Article 4(3) of the directive under which Member States send copies of their national specifications to the Commission if they want them to be recognised; both the Commission and other Member States take action, following a procedure laid down in the directive, if they wish to have approval for any such technical specifications to be withdrawn. Since there are already large numbers of national specifications in existence this form of technical specification will predominate for some time. As harmonised standards are developed, and European technical approvals are issued, the national specifications will gradually be replaced.

Harmonised standards

Harmonised standards are produced by the European standards organisations, CEN and CENELEC, generally based on mandates for their development issued by the Commission; such standards may also originate from a CEN initiative. Alternatively, national standards organisations may apply to CEN or CENELEC to have one of their standards adopted as a harmonised standard, or for a mandate from CEN or CENELEC to develop a standard. CEN or CENELEC have few funds available for developing standards on their own or national initiatives with the result that most harmonised standards will result from a Commission mandate, under which the Commission provides funds. There are at this moment no harmonised standards for construction products, although a draft guideline for mandates has been issued. This outlines two main types of standard – category A, covering design and workmanship, and category B, relating to products. The Eurocodes (see p 158) could fit into the former category.

Representatives of the national organisations meet in a variety of technical committees (TCs) to develop European harmonised standards; reporting to the technical committees are the project teams who do most of the drafting work. Under CEN rules, European standards are issued for a three year period as ENV (European norms for validation) for validation and assessment. At the end of this period there is a two year period for redrafting. Once this is complete and the final version adopted in CEN or CENELEC there is a short period of about three months for the equivalent national standards to be withdrawn. For Eurocodes, however, (see p 158) the period for withdrawal of the national code is not fixed and may be extended to up to two years or so.

The standards issued by the national organisations, worded identically with the harmonised standard, will have two or even three numbers – the national standard number such as a BS or DIN number (DIN is the German standardisation body), the European standard number, and possibly the International Standards Organisation number. Thus BS 5750, dealing with quality systems, is worded the same as European EN 29000 series and international ISO 9000 series (note that BS 5750 has parts while European and international standards use numbers for the same purpose e.g. EN 29001, 29002, etc.).

European technical approvals

A European technical approval is a favourable assessment of the fitness for use of a product for an intended use, based on fulfilment of the essential requirements for construction works for which the product is used and is based on examinations and tests of the construction product.

Thus, European technical approvals provide a means whereby products, for which there are no specified harmonised European standards or approved national standards, or which differ significantly from an existing standard (especially those of innovative nature), can be proved to meet the essential requirements. Approvals are to be issued by bodies authorised by a Member State, generally for a period of five years, on the basis of performance against the requirements of the relevant interpretative documents and common guidelines. There is, however, a long way to go before European technical approvals can be made available, although the rate of production should be much faster than for European standards.

European technical approvals will be similar to agrément certificates currently issued in the UK by the British Board of Agrément (BBA), by the CSTB (Centre Scientifique et Technique du Bâtiment) in France and by the BAM (Bundesanstalt für Materialforschung und Prüfung) in Germany. The development of a European level for technical approvals has required the creation of the European Organisation for Technical Approvals (EOTA) which was set up in October 1990. This organisation will oversee the award of European technical approvals. There are close links between the EOTA and the European body for agrément organisations – the Union européenne pour l'agrément technique dans la construction (UEATC). National approval bodies are designated by Member States, which are required to notify other Member States and the Commission of the bodies which they have authorised to award European technical approvals. Until the introduction of the European technical approvals all mutual recognition of agrément certificates was based on bilateral agreements between members of the UEATC. Bilateral agreements were not comprehensive and the situation was clearly unsatisfactory.

Attestation of conformity

The responsibility for the attestation that products are in conformity with the requirements of a technical specification lies with the manufacturer or his agent in the Community, if the product is imported.

However, under the Construction Products Directive products specified by the Commission as playing a minor role with respect to health and safety can be marketed on the basis of a declaration of compliance with the 'acknowledged rule of technology' by the manufacturer. The Commission is due to draw up and manage a list of such products. There is no particular compliance requirement before the declaration is made. Such products cannot bear the CE mark.

If construction products, other than those on the list, are to bear the CE mark, they must undergo a conformity assessment and attestation. Attestation requires the manufacturer to have a 'factory production control system to ensure that production conforms with the relevant technical specifications'. The directive provides four routes to attestation of conformity. The choice of route is by the Commission, assisted by the Standing Committee (see below), according to how vital the product is to the meeting of the essential requirements, the variability of the products etc. The manufacturer has no choice. The four routes are:

(1) *Certification by an approved certification body* that the product conforms (both the manufacturer and approved body have specified tasks).
(2) *Declaration by the manufacturer* that the product conforms. This requires one of the following:
   (a) initial type-testing of the product, factory production control and possibly further quality control testing procedures, *with an approved body certifying* the factory production control system and possibly carrying out audits (known as 'quality assurance' in the UK); or
   (b) *initial type-testing by an approved laboratory* and factory production control; or
   (c) *initial type-testing by the manufacturer* and factory production control.

Approved bodies

Member States must provide the Commission with a list of names and addresses of certification bodies, inspection bodies and testing laboratories designated for the purposes of the Construction Products Directive.

Standing committee on construction

The Construction Products Directive provides for a standing committee which is composed of two representatives of each Member State, who may be accompanied by experts. The function of the committee is to advise the Commission on a variety of matters connected with the directive, including:

(1) The drawing up and acceptance of interpretative documents.
(2) The compliance of national technical specifications, harmonised standards and European technical approvals with the essential requirements.
(3) The listing of products which are said to have a minor role regarding health and safety.
(4) Standardisation mandates and guidelines for European technical approvals.
(5) The procedures to be introduced for the attestation of conformity of products.

Structural Eurocodes

Much work has been undertaken on the development of a series of structural Eurocodes. These are both 'technical specifications' and 'category A standards'. The Eurocodes are concerned with the design and execution of construction works, including the installation of construction products. In many product areas the development of product standards is linked to the development of the design codes and will follow on once the Eurocodes are finalised. Reportedly, about half the Commission funds for standardisation are earmarked for development of the Eurocodes because of their pivotal position in relation to product standards. The nine Eurocodes being developed relate to the essential requirements, particularly mechanical resistance and stability, and safety in case of fire. The Eurocodes are being developed by 54 project teams under the guidance of the CEN technical committee TC 250. They are numbered as follows:

EC1 Actions (i.e. loads) – this incorporates the former EC1, which dealt with the basis of design etc., as Part 1
EC2 Design of Concrete Structures
EC3 Design of Steel Structures
EC4 Composite Structures
EC5 Timber Structures
EC6 Masonry
EC7 Foundations
EC8 Structures in Seismic Regions
EC9 Aluminium structures

The development of the Eurocodes is leading to much co-operative research, often required to validate the basis of the design codes and make cost comparisons, and other collaborative ventures. One example of collaboration, making use of funds from the European Commission under the COMETT programme, is a European-wide project to provide a comprehensive set of teaching materials based on Eurocode 3 for undergraduate, post-graduate and continuing professional development use. The cross-border contacts generally involved appear to be be stimulating innovation and new approaches. It also seems that some of the eastern European countries are recognising the benefits of European harmonisation and are moving to the Eurocodes (and other European standards) rather than update their own codes. It may even be that unless certain countries move rapidly, including the UK, where degree courses are relatively short and educational resources limited, others will be better placed to take advantage of the opportunities.

The Eurocodes are intended to be comprehensive documents and have made few sacrifices to simplicity of use. However, simplified documents, such as 'the essentials of EC3', and computer software packages are being developed which will provide for ease of use. Amongst the studies to be undertaken are cost comparisons between design carried out to the existing national design codes and the Eurocodes. The full transition to Eurocodes is likely to take some time. One British engineer closely involved with the development of EC3, which is at a more advanced state of development than the majority of the Eurocodes, estimates that it will not fully replace BS 5950 until the year 2000.

Testing and certification

Compliance with standards and specifications is a major concern of producers, suppliers, purchasers and users of goods. To ensure a given level of quality, which may be required by law, testing, certification and inspection may have to be undertaken. At present, goods often have to be retested and rechecked at considerable expense when they cross national borders. The Commission is determined to rectify this.

Testing and approval of test laboratories

Testing is a technical process using specified procedures to measure particular characteristics of products. Testing may be carried out by independent laboratories at the request of manufacturers. There is a huge volume of testing carried out annually in the Community, with the UK alone spending an estimated £20 billion. The approval (in the UK referred to as accreditation) of a test laboratory is a formal acknowledgement of its competence, the quality of equipment and its quality management system, through expert technical assessment based on agreed criteria, with periodic checking and reassessment.

In France, laboratory approval is carried out by the Réseau National des Essais (RNE) and in the UK by its counterpart, the National Measurement Accreditation Service (NAMAS). The concept of the approval of test laboratories is gaining ground in Europe with eight out of 12 Member States having systems in place or being developed. However, mutual recognition of tests results is still not available throughout the Community. Presently, a network of bilateral agreements is being developed between the national laboratory approval bodies.

The introduction of the EN 45000 series of European standards (worded as the BS 7500 series) for test laboratories and their accreditation should facilitate harmonisation.

Quality assurance

In the UK a system of quality assurance for firms has been developed and is gaining widespread use in the construction industry, with many clients, including public clients, insisting on contractors who have a certificate of quality assurance issued by an accredited body. The system has also spread to consultants and other service organisations. Other Member States have or are introducing equivalent systems. A characteristic of the UK system is that the certification bodies are all private organisations, such as Lloyd's Register Quality Assurance, Bureau Veritas, BSI Quality Assurance and a great variety of trade based organisations such as CARES (the UK Certification Authority for Reinforcing Steels, which deals with the suppliers of reinforcing steel). Certification bodies may be accredited in the UK by the National Accreditation Council for Certification Bodies (NACCB), but do not have to be accredited to function. They are accredited only for specific activities and may also carry out non-accredited activities.

In the UK the assessment of quality systems is usually undertaken against the requirements of BS 5750 (worded as the EN 29000 and ISO 9000 series) which relates to management structure, responsibilities, training and quality procedures and records.

If implemented properly, quality assurance (QA) should result in a higher and more consistent quality of service or product and reduced wastage and reworking costs and reduced exposure to product liability claims. There are serious dangers, however, that QA can become a costly and bureaucratic system which is imposed on a reluctant line management, who may see it as an irrelevance.

Mutual recognition

Mutual recognition of national quality assurance certificates or of one-stop approval for certification bodies is still some way from reality, although there are already some bilateral arrangements.

National influence on the development of standards

There are some who claim that CEN and CENELEC technical committees are being used by the larger Member States, particularly France and Germany, through trade and other bodies, to bias the drafting in favour of their national industrial strategy and in the interests of their national manufacturers. German companies are claimed to put money and effort into their trade associations as a means of influencing the development of standards. It appears, certainly, that German and French manufacturers are more willing to devote the considerable resources needed to support the work of the committees, so that the results suit their industries. Another factor is that both the Germans and, to a lesser extent, the French, tend to have built up their views and approaches on an impressive body of research and other data. Succeeding with contrary views is difficult without equally valid evidence, although there appears to be a pragmatic willingness to base the development of standards on reason and the rule of technology, provided the end results are not compromised. Whether manufacturers and others in the other Member States realise the significance of the developments is not certain.

The Commission Green Paper on the development of standardisation

In October 1990 the Commission issued the Green Paper on the development of European standardisation [9.7]. The Green Paper aimed to stimulate producers and users of standards to become more involved in the development of standards, and to put forward for discussion a number of new ideas to speed up the process. Various recommendations for procedures and processes in relation to the development of new standards were proposed. These include the use of 'drafting secretariats' or project teams to work on the drafting of documents, the development of more associated standardisation bodies by CEN and CENELEC and shorter enquiry period and response times for comments. The direct applicability of European standards, without waiting for national transposition, is also proposed. Of particular importance is the modular approach to attestation, which has now been added to the 'new approach' [9.8]. All new directives will incorporate this approach to attestation but existing directives will only be affected when revised.

A new organisational structure is also proposed, with a European Standardisation Council at the top, which would include EFTA and EC representation amongst others, and a European Standardisation Board reporting to the Council and made up of representatives of the European standardisation bodies. This body would undertake the management and co-ordination of the standardisation programme. European and national standardisation bodies would of course remain part of the new system, probably undertaking much the same roles as at present.

Conclusion

Companies related to the construction industry will be affected by the Construction Products Directive, new arrangements on testing and certification and the Eurocodes. Clients will have a wider choice of product and the business environment will become tougher for manufacturers and designers. However, it will be several years before all supporting European harmonised standards or the Eurocodes are in place. The transitional period, with a mix of national and European specifications, will create complications and problems, but as ever opportunities will be created and those with the expertise will be able to exploit them. Unfortunately, although the new approach has succeeded in accelerating the standardisation process, it does little directly to reduce the unnecessary variety of products available in European markets. For example, the electrical sockets normally used in the UK and France may both meet the essential requirements of health and safety, but it is not convenient to have a mixture in one house.

Chapter 10

Standardisation on Liabilities and Other Matters

This chapter examines directives, proposals and policy in relation to liability for construction, warranties, compulsory insurances, construction procedures, roles and forms of contract. It also considers liability for products and damage caused by waste.

The liability of designers, developers, suppliers and contractors is a serious concern for those in the construction industry and requires costly insurances. In the UK the situation has been confused by recent and substantial changes in the case law relating to liability in tort, with cases such as *D&F Estates Ltd and others* v. *The Church Commissioners for England and others* [10.1], *Greater Nottingham Co-op* v. *Cementation* [10.2] and *Murphy* v. *Brentwood District Council* [10.3]. The consumer also has a clear interest. Under the French system, where a mandatory 10 year warranty and insurance is provided, the building owner appears to benefit. A claim can be made against the insurer; compensation is rapid and the relative liability of the designer and contractors involved is established subsequently by the insurer (the system is described in chapter 13 in the section dealing with France).

Commission policy in this area has two strands. The first is concerned with consumer protection, for example the product liability directive (1985) (see p 164). A proposal at the end of 1990 (see p 166) deals with the liability of suppliers of services including construction. The second strand is concerned with reducing the barriers introduced for construction firms through having to deal with differing systems and liabilities when undertaking work in other Member States; this could result in proposals harmonising the division of responsibility in construction projects, forms of contract and other aspects.

For both strands Article 100a, the article dealing with the approximation, or harmonisation, of laws relating to the establishment and functioning of the internal market, is the legal basis for action. This article envisages a high level of protection in relation to health, safety, environmental protection and consumer protection.

Harmonisation on product liability; introduction of strict liability

The product liability directive 85/374/EEC [10.4] was adopted in 1985 and was implemented in the UK by the Consumer Protection Act 1987. It attracted relatively little comment in the construction industry press. The directive applied to 'movables' including products incorporated into 'immovables'. Buildings and civil engineering works themselves are not therefore covered by the directive itself, while aggregrates, pre-cast concrete beams and other components clearly are. The directive specified that Member States 'may provide that a producer's total liability for damage resulting from a death or personal injury and caused by identical items with the same defect shall be limited to an amount which may be not less than 70 million ECU'. While the supply of buildings was excluded under the directive, Part 1 of the Consumer Protection Act provides for liability of suppliers and others provided the supply was not 'effected by the creation of or disposal of an interest in the land'. Thus, a contractor in the UK who constructed a building or work on another's property could be liable for personal injury or damage to personal property caused by a defect in the work.

Strict liability

The directive introduced strict liability for the producer whose defective product caused personal injury or damage to personal property, and created an offence of supplying goods which contravene the general safety requirement. With strict liability the producer is liable whether or not there has been negligence. Before the implementation of the directive strict liability in the UK was restricted to very few areas of the law, one of the prime examples being the *Rylands* v. *Fletcher* [10.5] principle. Under this principle the owner of land is strictly liable if something which he has brought onto or collected on his property escapes and causes injury or damage. Defences against liability are specified in the product liability directive, but do *not* include that the product was designed by someone else; thus a contractor who constructed a building to the design of an architect could be liable. It is possible that designers could also be liable under the directive (or its implementation). According to the London law firm Titmuss, Sainer and Webb quoted in *Building* [10.6] 'the product role of architects, engineers and other professionals primarily relates to the provision of design and supporting calculations. As such they would be "producers" and persons who, by putting their names to the product hold themselves out as being the producers of it'. There appears to be no other authority for this view.

Joint and several liability

The directive incorporated the rule of joint and several liability, which means that any of those sharing liability could be sued for the total amount of damages involved, even if they cannot recover any share from others who are liable, for example through liquidation. A small mistake which contributes to a major disaster could render the defaulter liable for all the consequences. This principle has been followed in the subsequent proposals on waste damage liability and liability for services discussed below.

Waste damage liability

The principles of strict, and joint and several liability have also been incorporated into the proposal for a directive on waste damage liability [10.7]. This has alarmed banks and others because if land is in their control, for example after the default on a secured loan, they could find themselves liable 'under civil law for damage and injury to the environment caused by the waste, irrespective of fault . . .'. Derek Wheatly, writing in the *Financial Times* [10.8], draws lessons from similar legislation which has been in force in the USA since 1980 – Comprehensive Environmental Response, Compensation and Liability Act (CERCLA) 1980.

Mathurin report

The different national practices in relation to defects in immovables have given rise to much debate about further harmonisation measures. In February 1990 the Commission services produced an internal consultative document [10.9] which they had commissioned Mr Claude Mathurin, a French *Ingénieur général des Ponts et Chaussées* (road and bridge engineer), to produce. This included analysis of practice in the Member States and sought opinion from the industry and others about the desirability of harmonisation.

The variations in national practice on liability are noteworthy. In Germany 'the guarantee of the builder or vendor of a new building is for five years after acceptance, but that of the contractor can be reduced by contract to two years', while in the case of gross negligence damages and interest can be recovered for up to 30 years. The UK, with its confusing mixture of liability in contract and tort, is referred to. In contract, liability extends for up to 12 years from a breach of contract, if the contract was executed in a certain way. In tort, case law (some key cases see p 163) has brought dramatic fluctuations in liability, only partially resolved by the Latent Damage Act 1986. In France, liability lasts for 10 years and must be covered by insurance.

Subsequent attention has been focussed on the liability, guarantees and insurance aspects of the report, but Mr Mathurin recommended further discussion on three potential directives in the first instance. The first would deal with the certification of compliance with the essential requirements for buildings, the roles of the client, principal designer and contractor, and external technical inspection amongst other aspects. The second would deal with the tasks of designers, including remuneration, drafting of technical specifications and design contracts, while the third would be devoted to liability, guarantees and insurance. Acceptance of works, specific liability of the builder or vendor, guarantees, housing insurance, control insurance and sub-contracting were to be in the third directive. Amongst those consulted, architects and engineers were generally in favour of harmonisation, contractors and manufacturers were split, insurers and inspectors were unanimously in favour of the suggestions, while government administrations were cautious and favoured setting up a committee. Subsequently FIEC, the European contractors federation, has agreed that there is a case for a limited degree of harmonisation.

In a subseqent meeting of an ad-hoc committee of Member States representatives, referred to as GRIM (in curious franglais Groupe Réglementation Information et Management), six of the Member States, including the UK, were opposed to harmonisation and held that the Commission had failed to prove that the differences in liability laws were a barrier to trade or that harmonisation was needed in the interests of consumer protection. Further study was recommended.

Proposal covering liability for construction services

The Commission, in the form of the consumer protection directorate, surprised the industry in January 1991 by including design and construction of immovable property within the scope of a proposal for a directive on the liability of suppliers of services [10.10]. The proposal covers most services including health care, transport, playgrounds, hotel and leisure services, as well as services in relation to building and civil engineering works. It does not apply to package travel or to waste services (see p 165). Also it does not apply to services provided by one individual to another. The explanatory memorandum to the proposal explains that safety is an essential feature of the internal market and that to provide for freedom of movement consumer confidence must be increased. It contrasts the situation in Italy and the UK where the consumer who has been harmed must generally prove the supplier to be at fault, whereas in Spain, Denmark, Greece, Belgium and in some cases Germany, the burden of proof is reversed to the consumer's advantage; the memorandum goes on to assert that 'in Europe, all consumers and injured persons should have equal rights and their chances of receiving compensation should be based on standard principles'.

The proposal is rather less far-reaching than suggested in the press. It covers damage to persons and property and to financial material damage resulting directly from the physical or material damage, but not economic loss or indirect effects, such as suicide or the loss of an opportunity to win a contract. The important point is that damage to property is only covered if it was of a type normally intended for private use or consumption and intended for or used by the injured persons, or principally for his private use or consumption. This excludes much potential liability arising from design or construction work; damage or loss suffered by a company would not be covered, while injury to individuals and damage to private property, for example arising from a bridge collapse or negligent soils testing, would be. Damage to private property could include that to a private house, a private jetty or car (not a company car). Personal injury could cover disease; examples of possible causes are injury caused by falls of material, illness caused by materials used in construction or failure to allow for radon accumulating in a dwelling, and injury resulting from an explosion caused by failure to allow for methane build-up, possibly from natural sources or waste.

Liability under the proposal only arises when there has been negligence and is not therefore correctly described as 'strict'; but the burden of proof is reversed. This would mean that the supplier of the service must demonstrate absence of negligence. The injured person must merely provide proof of the damage and the causal link between the damage and the allegedly negligent performance of the service. Joint and several liability is specified.

Liability periods and the longstop on liability are of great concern, particularly in the UK where the Latent Damage Act 1986 was introduced to set limits. The proposal incorporates special limits applying to the design or construction of immovable property; liability would last for 10 years from the day on which the plaintiff became aware or should reasonably have become aware of the damage, the service and the identity of the supplier of the service. This compares with the situation in the UK where, under the Latent Damage Act, liability continues until six years from the date of the breach or, if later, three years from the date of reasonable discoverability. The proposal specifies three years rather than 10 for services not connected with the design or construction of immovable property.

In relation to design or construction of immovable property the proposal states that all liability should cease 20 years from the date of the provision of the service which caused the damage. The longstop period proposed for services not related to immovable property is five years. The possibility of a 20 year period worries the industry because excellent records, sufficient to prove absence of negligence, would have to be maintained until there is no longer any possibility of a claim. Insurances would also have to be maintained, at some additional cost.

Since the legal basis for the proposal is Article 100a of the Treaty, qualified majority voting will apply. It is difficult at this stage to predict whether it will be adopted without much change. The date specified in the draft for implementation in Member States is 31 December 1992.

The proposed directive deals with only one specific aspect of construction procedure and liability. This still leaves the possibility for further initiatives on the Mathurin proposals. Mr Caronna of DGIII of the Commission indicated in January 1991 that proposals would be developed dealing in turn with terminology, certification of construction firms to ensure minimum standards of performance, and finally consistent liability in relation to construction. The latter would deal with liability beyond that relating only to personal injury and damage to personal property.

Standard form of contract

Many parties besides Claude Mathurin would like to see the Commission propose a standard form of contract for the Community construction industry. One of the interested informal groups consists of a group of major contractors from around the Community with coordination being provided by McKenna & Co, a London firm of solicitors who specialise in construction law. It was also reported in *New Civil Engineer* on 31 January 1991 that an influential group of British engineers was to lobby the Commission to develop a standard form, possibly based on the New Engineering Contract recently developed by the Institution of Civil Engineers. However, the

Commission has to date responded by saying that it has other priorities to attend to before the matter of a standard form will be considered.

Chapter 11

Mutual Recognition of Qualifications

Professional qualifications are particularly important in the construction industry. In many countries there are legal requirements restricting certain activities to the holders of particular qualifications. In the UK there are very few such restrictions although the title 'architect' is protected by law. Even where there are no legal restrictions, qualifications may be required by employers as an indicatin of knowledge and competence. Thus, to have true freedom of movement for workers and to have freedom to provide services in the Community professional qualifications awarded in other Member States must be recognised. Article 57 of the Treaty, to which important additions were made by the Single European Act, is the legal basis for measures in this area. The White Paper on the Completion of the Internal Market [11.1] listed a number of proposals which were to be produced.

Directives providing for mutual recognition of qualifications have been adopted in relation to many professions, including architecture and engineering. This chapter explains how the approach to mutual recognition has changed and describes the main provisions relating to architects, who have their own directive, and to engineers who are covered by a general directive. The draft for a second general directive relating to qualifications for which a degree or equivalent is not required is also referred to.

The original approach to the mutual recognition of qualifications was called the sectoral approach, and involved detailed harmonisation of the minimum criteria which the training for a particular profession had to meet. Any person who held the qualification would have had a similar background of education and training in a particular discipline. Since the education and training had been harmonised the qualifications would be equivalent and would be recognised in all the Member States. Qualifications for medical practitioners, nurses responsible for general care, dental practitioners, veterinary surgeons, midwives, pharmacists and general medical practitioners were harmonised in this way. Medical practitioners were the first with a directive in 1975.

Architects

The architects' directive 85/384/EEC [11.2] followed the sectoral approach, but instead of specifying criteria it specified the areas to be covered during studies.

> '. . . studies shall be balanced between the theoretical and practical aspects of architectural training and shall ensure the acquisition of:
> (1) An ability to create architectural designs that satisfy both aesthetic and technical requirements.
> (2) An adequate knowledge of history and theories of architecture and the related arts, technologies and human sciences.
> (3) A knowledge of the fine arts as an influence on the quality of architectural design.
> (4) An adequate knowledge of urban design, planning and the skills involved in the planning process.
> (5) An understanding of the relationship between people and buildings, and between buildings and their environment, and of the need to relate buildings and the spaces between them to human needs and scale.
> (6) An understanding of the profession of architecture and the role of the architect in society, in particular in preparing briefs that take account of social factors.
> (7) An understanding of the methods of investigation and preparation of the brief for a design project.
> (8) An understanding of the structural design, constructional and engineering problems associated with building design.
> (9) An adequate knowledge of physical problems and technologies and of the function of buildings so as to provide them with internal conditions of comfort and protection against climate.
> (10) The necessary design skills to meet building users' requirements within the constraints imposed by cost factors and building regulations.
> (11) An adequate knowledge of the industries, organisations, regulations and procedures involved in translating design concepts into buildings and integrating plans into overall planning.'

There are considerable differences in architectural 'formation' around the Community, particularly in the area of structural engineering and design. Under the directive acquiring 'understanding' is clearly a higher objective than acquiring 'an adequate knowledge' and applies to structural design, the constructional and engineering problems associated with building design amongst other areas. It seems doubtful that architectural education in all the Member States currently meets this requirement. Item 10, which requires an ability to design within cost constraints, is another area where architects in some Member States may be weak. In the UK, for example, the cost aspects of design are generally handled by the quantity surveyor, a professional not part of the traditional systems in mainland Europe. In consequence, trainee architects in the UK may receive less instruction related to cost than those in other Member States.

Apart from stipulations on content, the architects' directive specifies that the minimum full-time training for recognition shall be four years at a university or equivalent, or six years of study, of which at least three years must be full-time.

There have been problems in implementing the architects' directive in some Members States. In Germany it was reported in November 1990 that only one of the 11 *Länder* of the former west Germany had implemented the directive, though informal arrangements had been made; Ireland had had problems, and Spain had written to the Commission attempting to justify restrictions it was imposing on incoming architects. The Commission was not sympathetic to Spain's case. At the time of writing formal action is being taken by the Commission against Greece for non-compliance, and there are queries outstanding with Belgium and Spain.

The range of education, professional training, and roles undertaken by architects across the Community lead to considerable problems in agreeing the directive. In Germany, the education and function of architects encompasses what in the UK is generally done by structural engineers, and to some extent the contractor's site management, since much contracting is done on a trades contract basis with organisation and co-ordination carried out by the architect. It took 17 years to reach final adoption of the directive, which finally took place in 1985. Implementation of most provisions was to be by August 1987, and the remainder a year later.

The first general directive on mutual recognition of qualifications

The problems with the architects' and other directives in agreeing training criteria, or in specifying course content, led the Commission to adopt a new approach, one which simply requires the Member States to recognise the qualification of someone from another Member State, with certain safeguards.

The first such directive was adopted by the Council in 1988 – the directive 'on a general system for the recognition of higher-education diplomas awarded on completion of professional education and training of at least three years duration' [11.3]. It came into force on 4 January 1991 and applies to all professions which are regulated in some way by the state and which require at least three years' university level education or equivalent, and in many cases experience in the profession. Under it professional engineers, surveyors, town planners, builders, teachers, accountants, lawyers and others may apply to the body which regulates the profession in another Member State to recognise their qualification.

In most Member States many of these professions are regulated by law, whereas in the UK, Ireland and certain other Member States, the professions are self-regulating. The directive is aimed at professions which are state regulated, but the different circumstances in the UK and elsewhere have been dealt with by deeming an activity to be regulated 'if it is pursued by the members of an association or organisation the purpose of which is, in particular, to promote and maintain a high standard in the professional field concerned and which, to achieve that purpose, is recognised in a special form by a Member State etc.'. The chartered professional bodies in the UK fulfil the criteria and have been given delegated power to exercise functions under the directive.

The body which regulates the profession must accept a qualified person from another Member State, subject only to certain requirements which may be imposed. Where a person's professional experience is at least a year less than that normally required for qualification in the host state, additional experience may be required. The other possibility is that the applicant may be required to complete an adaptation period of up to three years or take an aptitude test. Once the qualification has been recognised the equivalent qualification in the host Member State may be used, though this may involve becoming a member of an organisation or association, as in the case of becoming a Member of the Institution of Civil Engineers in the UK.

There are difficulties for professions which do not exist in the same form in other Member States. UK quantity surveyors are broadly equivalent to construction economists, but building surveyors who often carry out design work in the UK have no direct equivalent in most other Member States. They have been concerned about being unable to carry out design in Member States where that is restricted to architects or engineers; in Belgium, for example, only registered architects may carry out design involving the outside of a building, even if it merely involves an extension.

Implications

There are, in some cases, considerable differences in the educational and training requirements leading to qualifications which are mutually recognised. This may lead to shopping around for the quickest, cheapest or easiest way to achieve a qualification. For example, a French would-be engineer finding the route blocked in France may be able to obtain a qualification in the UK which will then have to be recognised. David Waller writing in the *Financial Times* [11.4] quotes differences between the route to equivalent accountancy qualifications in the UK, where three years post degree are required, and Germany, where five years are required, and the degree must normally be in economics. 'From January (1991), Germans will be able to become chartered accountants and then switch to being fully-fledged *Wirtschaftsprüfers*. Tests will still have to be undergone, but they will not be as severe as the normal path to Institut der Wirtschaftsprüfers.' If many people start shopping around for qualifications it could force the professional and regulatory bodies around the EC to bring their educational and training requirements closer together.

A draft second general directive

A second general directive, similar to the first but dealing with regulated qualifications which do not require a degree or equivalent, is at an early stage of development at the time of writing. In the UK, the Department of Employment is handling the consultative processes. One problem is that a wide range of activities which are regulated in other countries are not regulated in the UK. For example, in certain Member States it is necessary to have a qualification to practise as an electrician, whereas in the UK no qualification is needed.

Chapter 12

Structural Funds and Funds for Research and Development

This chapter describes the ways in which the Community provides funds which may benefit the construction industry. The funds include those for infrastructure development, particularly in the poorer Member States. The sums of money available are considerable and offer opportunities for providing consultancy to authorities on the preparation of general development plans which the Commission requires in connection with applications for grants. Once the regions have been awarded funds there are opportunities for winning contracts in design and construction. Contract award procedures for projects supported by the funds must follow the public procurement rules, and should therefore be advertised in the *Official Journal* and in *Tenders Electronic Daily* (TED). Consultants should be aware of the funds as a source of finance for their clients. Contractors may be able to obtain funds in connection with infrastructure projects or for tourism or other development projects. The funds may provide finance for developers when certain types of project are involved. The funds are also important since the projects supported can have a substantial economic impact on the regions involved, as indeed they are intended to. This can produce a knock-on effect on other development, such as commercial and industrial development, bringing further spending on construction.

This chapter also briefly describes the financial assistance the Community makes available for cross-border collaborative research and development. These funds are targeted on specific areas of science and technology, some of which should be of interest to construction firms and others connected with the construction industry.

European Social Fund

The Community redistributes funds from the richer regions of the Community to the poorer regions through a number of funds collectively referred to as the 'structural funds' since their aim is to bring about change to the economic structure of the target regions. The original such fund was the European Social Fund (Treaty Articles 123-128). Its purpose is to 'improve employment opportunities for workers in the common market and to contribute to raising the standard of living, . . . it shall have the task of rendering the employment of workers easier and of increasing their geographical and occupational mobility within the Community.' The fund is used to finance training schemes and vocational guidance, to provide subsidies for recruitment into newly created long-term jobs and to help people set themselves up in self-employment. Britain receives more grants than any other Member State excluding Italy under the European Social Fund.

European Regional Development Fund

The European Regional Development Fund (ERDF) was first set up in 1975. It is used to support capital intensive projects that contribute to the economic development of a region, for example transportation, communications, energy, water, sewerage, land drainage, flood protection and tourist infrastructure projects. It does not cover housing, school or hospital construction. Under the reform of the structural funds (see below) the ERDF is now directed almost exclusively at the poorer regions of the Community.

Other instruments

The Community has other instruments available to promote structural change. These include the European Agricultural Guidance and Guarantee Fund, Guidance Section, certain funds under the European Coal and Steel Community Treaty, and loans available through the European Investment Bank (see p 179).

Reform of the structural funds

The Single European Act introduced a new Title – 'Economic and Social Cohesion' – into the Treaty of Rome. There were five new Articles – 130a to 130e – under this title. Article 130a required the Community to 'develop and pursue its actions leading to the strengthening of its economic and social cohesion' and to 'aim at reducing regional disparities between the various regions and the backwardness of the least-favoured regions'. Article 130c incorporated the European Regional Development Fund into the Treaty and set out its objective as to 'help redress the principal regional imbalances in the Community through participating in the development and structural adjustment of regions whose development is lagging behind and in the conversion of declining industrial regions'. Article 130d required the reform of the funds which are collectively referred to as the 'structural funds (European Agricultural Guidance and Guarantee Fund, Guidance Section, European Social Fund, European Regional Development Fund)'. The reforms were achieved through four Council Regulations adopted in 1988 [12.1] and came into effect on 1 January 1989.

Under the reforms funding was doubled to ECU 20 billion, in 1988 ECU 7.7 billion of which was committed to grants under the structural funds. For comparison, Community spending on agricultural market support (approximately 60 per cent of the EC budget) was ECU 27 billion in 1988. By 1992 the figure for grants under the structural funds is to be increased to ECU 14 billion (at 1988 prices). Most of the new money is directed into the European Regional Development Fund which is targeted at the poorer regions of the Community (see 'objective 1' below).

Under the reforms the objectives of the funds are:

Objective 1: Promoting the development and adjustment of the regions whose development is lagging behind (i.e. where per capita GDP is less than, or close to, 75 per cent of the Community average). Eighty per cent of the European Regional Development Fund is directed at these regions which are Ireland, Northern Ireland, Portugal, most of Spain excluding the Madrid region and northern parts of the country, Corsica, Sardinia, southern Italy including Sicily, and all of Greece.

Objective 2: Converting the regions, frontier regions or parts of the regions seriously affected by industrial decline.

Objective 3: Combating long-term unemployment.

Objective 4: Facilitating the occupational integration of young people.

Objective 5: (a) adapting production, processing and marketing structures in agriculture and forestry;
(b) promoting the development of rural areas.

Under objective 1 the main beneficiaries are Spain (32.6 per cent), Italy (24.5 per cent), Portugal (17.5 per cent) and Greece (16.2 per cent). In addition Ireland gets 5.4 per cent, France 2.1 per cent and the UK 1.7 per cent. Under objective 2 the UK gets the largest slice at 38.3 per cent, Spain gets 20.7 per cent, and France gets 18.3 per cent. Most of the European Regional Development Fund is allocated to the Member States under programmes agreed with the Commission, but 15 per cent of the fund is reserved to be allocated by the Commission for projects which will benefit the Community overall.

European Regional Development Fund grants, European Investment Bank loans, as well as European Coal and Steel Community loans, are available for 'investment in public infrastructures of economic importance: roads, telecommunications, airports, ports, inland waterways, navigation, dams, vocational training centres, facilities for transport operators, tourist facilities, advisory and support services for business'. Water supply and sewage treatment facilities are also included. The normal maximum assistance is 50 per cent of the total cost, but for the regions covered by objective 1 it can be up to 75 per cent.

Planning and administration of the funds

The overall co-ordination of the funds is handled by DG XXII, while administration in connection with objectives 1 and 2 is handled by DG XVI, in connection with objectives 3 and 4 is handled by DG V, and DG VI (agriculture) and DG XIV (fisheries) are involved with objective 5. National and regional authorities are required to draw up rolling development plans covering three to five years. These are agreed with the Commission and include programmes of development to be supported by the funds. The first plans were prepared in 1989. The plans provide information which can be valuable for those considering regions of the Community in which to concentrate, either with a view to winning contracts or in which to invest. Applications for grants are handled by the national agencies, not by the Commission. In the UK the contact point is the World Aid Section, part of the Projects and Exports Policy Division of the Department of Trade and Industry [12.2]. Community resources are intended to complement those awarded by Member States.

Special programmes aligned to the structural funds

Special programmes include the Integrated Mediterranean Programmes which cover Greece, Italy and France. These can be used for infrastructure and environmental projects. Another programme called PEDIP is aimed at developing Portuguese industry.

Fund to improve road and rail links

Another, relatively modest, fund adopted in late 1990 is targeted at transport projects with a clear European dimension. It is intended to be used for combined road and rail links, bottle-necks which affect cross-border traffic and projects which help integrate peripheral regions into the Community. Only ECU 60 million in 1990 and ECU 118 million for 1991 were made available under this scheme which is administered by DG VII (transport). These amounts are additional to the European Regional Development Fund and other sources of finance.

European Investment Bank

The European Investment Bank (EIB) grants loans and gives guarantees which facilitate the financing of:

'(a) projects for developing less-developed regions;
(b) projects for modernising or converting undertakings or for developing fresh activities called for by the progressive establishment of the common market, where these projects are of such a size or nature that they cannot be entirely financed by the various means available to individual Member States;
(c) projects of common interest to several Member States which are of such a size or nature that they cannot be entirely financed by the various means available to the individual Member States.' [Treaty Article 130]

In practice about 60 per cent (about ECU 5 billion) of the bank's resources go into the objective 1 regions. The bank supports infrastructure projects and projects concerned with the modernisation or conversion of firms or with the creation of new activities. The major UK borrowers from the bank have included Electricity Boards, Water Boards, Local Authorities (particularly Scottish Local Authorities), British Nuclear Fuels Ltd, postal and telephone services, gas services, British Rail, oil and chemical companies and other private firms.

The low interest rate on EIB loans is the main advantage. The problem is the cumbersome application and administration procedures, although these have been simplified and shortened as part of the reforms of the structural funds.

In 1989 the EIB lent 12.2 billion ECU, 11.6 billion of which was inside the Community. In its 32 years of existence the bank has only had one failed loan, a 300 000 ECU loan for a German hotel project. Until the establishment of the European Bank for Reconstruction and Development it was required to start activities in eastern Europe, and lent 1 billion ECU to Poland and Hungary. In Poland the money was lent for railway construction and gas networks, while that in Hungary went for development of telecommunications and the electricity industry.

Impact of the structural funds

Before the reform of the funds it seems that in the UK, at least, many projects funded by the structural funds would have been funded in any case. The Commissioner with responsibility for the European Regional Development Fund was ironically once responsible for obtaining funds as Secretary of State for Scotland. He is quoted as saying:

> 'Every quarter we drew up a list of projects or companies due to get national assistance. We knew roughly what the UK as a whole and Scotland would get from the EC each year. So we just picked out as many projects as were needed to make up the UK quota, and sent the list off to Brussels. Back came the EC money, and the Treasury simply lopped that amount off its expenditure.'

With the reforms in the procedures, the increases in the funds and their targeting on the poorer countries of the Community, it is probable that the structural funds will achieve greater impact. It is claimed that the funds encourage fresh thinking. According to Chris Brewster and Paul Teague [12.3] the system resulted in, for example, new instruments to counter unemployment.

Funds for eastern Europe

The Community has a vested, as well as humanitarian interest, in securing the development of eastern Europe. The 'peace dividend' which has come about from the end of the cold war and various arms limitation treaties with the Soviet Union has resulted in a considerable reduction in spending on arms and defence. Some of these funds can be made available for development of eastern Europe. If the region fails to progress economically it could once more fall under totalitarian regimes and there could be a return to mutual hostility and increased defence spending. Unless the political changes bring quick returns to the peoples of these countries the Community could be facing another problem – of refugees and immigration. Some have estimated that as many as four million people from eastern Europe could try to move to Community countries. On the other hand, if the region flourishes it could be a valuable new market. It is in the Community's interest to provide support, and that support must bring results within the next four years or so, by which time most of the newly democratic countries of the east will be due for elections.

The Community has co-ordinated the policies of the Member States in this area and has made various grants and loans available. One programme set up in 1989 is the PHARE programme which provides aid to Poland, Hungary, Czechoslovakia, Bulgaria and Yugoslavia. It may also provide aid to Romania once the Community is satisfied that democracy has been established. Under the scheme, finance is made available for technical assistance, for training (related to trade) and for capital projects. Agricultural reform in Bulgaria, new statistical systems in Poland and a higher education exchange programme with Czechoslovakia are being supported. A high priority is being given to environmental measures.

Other projects which may interest construction companies are those concerned with water and waste treatment, measures to reduce pollution in the Baltic, installation of pollution monitoring equipment and training of local people. Because of the urgency the Commission is using accelerated approval procedures. Firms, institutions, federations, training centres and consultancies can become involved. Information can be obtained from the Commission Offices in Member States, Euro Info Centres and the Commission in Brussels [12.4].

Another source of funds for work in eastern Europe is the London based European Bank for Reconstruction and Development (EBRD) described in chapter 2.

Research and development funding

Large sums of money are available from the Community for research and development in support of a key objective of the single market – that of improving the competitiveness of Community industry. The funds are available to support pre-competitive research, i.e. '. . . only projects in a stage prior to industrial development, or not yet "market ready" [12.5]. Companies, as well as private or public research institutes and academic institutions, can apply, but projects are normally only 50 per cent funded from the Community research funds. Another requirement is that at least one partner from another

Member State is needed, since the funds also aim to promote cross-border co-operation. Organisations from EFTA countries can participate, but only in addition to partners from at least two EC countries. Small and medium-sized companies will benefit from special treatment and the Commission is keen to involve them.

Funding is organised in a series of overlapping 'framework programmes' lasting a number of years. The second framework programme spans 1987–91 and made available ECU 5.4 billion, while the Third Framework Programme (1990–94) is worth ECU 5.7 billion. Much of the funding is targeted on research information and communications technologies (38.9 per cent of the Third Framework Programme) and other areas outside the realms of the construction, but some of the programmes concerned with industrial and materials technologies, the environment and non-nuclear energies could be of interest to construction firms.

The following, selected from the extensive list of programmes, are of possible interest to firms associated with the construction industry (one cannot escape from acronyms in this sphere; they are usually derived from the title in English or French):

EURET is concerned with research into transport planning and traffic management, road safety, etc.
The **Raw Materials** programme, covers renewable raw materials such as wood products and waste recycling technologies including energy production from waste.
JOULE is concerned with models for energy and the environment, conservation in end-user sectors and renewable energies including geothermal energy.
STD aims to promote scientific co-operation with developing countries and covers conservation and better use of the environment amongst other aspects.
MAST is in the sphere of marine science and technology and covers coastal engineering.
NETT is a programme which aims to promote clean technologies by encouraging firms to make use of available research and consultancy.
STEP supports a number of aspects of environmental protection including water quality, soil and ground-water protection and protection and conservation of the European cultural heritage.

Applications for any of these funds require very thorough preparation. Information about the funds can be obtained from the Commission (DGXII) and Euro Info Centres. In any case, *EC Research Funding – A Guide for Applicants* [12.5] is essential reading. Another useful source of information in the UK is the Department of Trade and Industry's *Guide to European Community Research and Development Programmes* [12.6]. Information about current research projects and research results can be obtained using a variety of databases, most of which are free through the European Communities Host Organisation (ECHO) [12.7].

As well as research funds, there are Community funds supporting measures in education and training. **ERASMUS** is a programme to encourage movement of students between universities in different EC countries and joint study programmes. **COMETT** 'aims to create education partnerships between universities and business, and supports cross-border practical training in companies and the exchange of experts between universities and industrial firms'. **LINGUA** is aimed at improving language skills and supports exchange schemes for those between 16 and 25, especially as part of vocational training.

Chapter 13

Country Studies

The following country studies aim to provide an insight into the key features of the economy and likely development, the construction industry, business practices and other aspects relevant to undertaking work in each of the countries. They may help in drawing up a short-list of countries for further investigation, and to highlight the features of a country which might otherwise take one by surprise. Another function they should fulfil is to provide an insight into the great variations between the Member States of the Community. Community legislation may sometimes run counter to particular national traditions but it should be seen in the wider context. The studies cannot be comprehensive. Various sources of more detailed market information are available and in any case there is no substitute for visits to the countries of interest for fact finding and to make contacts. One should remember that procedures and structures in other countries are relatively easy to understand; the real problem is generally cultural, with different traditions and expectations colouring relationships and communication in subtle and unexpected ways. Sensitivity, openness and willingness to learn, not only about business but about all aspects of a country, are needed to work successfully with its nationals.

GERMANY

The momentous events of 1989, symbolised by the breaching of the Berlin wall, and subsequent reunification on 3 October 1990 have obscured the view of the state of Germany. Major restructuring proceeds, unemployment in the east threatens to soar to two million in a population of less than 17 million, the discount rate has climbed to a high (for Germany) rate of 6 per cent, but the effects of unification seem to be providing a cushion for the west against the recession which is biting elsewhere in Europe.

The constrast between the east and the west is dramatic. The large and powerful west German cars on the one hand and the cramped, polluting, two-stroke Trabant on the other provide a strong image for the contrast between the most prosperous and technologically advanced state in Europe and a country bankrupted and despoiled by a corrupt and ineffective system.

Germany, already the largest and most powerful Member State in the Community has become become even larger. Its population of 79 million (61.45 + 16.7 million) inhabitants is 40 per cent greater than Britain, the next most populous Member State. Can the Community accommodate such a predominant member?

Germany has a relatively short history as a single state, and borders have rarely been fixed. In 1815 the German states formed a confederation which left few powers with the federal 'diet' or Bundestag which was established in Frankfurt and which was unable to have direct authority over citizens. This broke up and in 1867 north Germany established a federal constitution which four years later became the basis of the German empire. In this the powers of the central state were limited. Even today the federal structure creates difficulties for the Community, when the federal authorities claim that they do not have the power to hand down certain provisions of EC law to the regional authorities.

Regional differences are expressed in culture, religion, traditions, lifestyle and temperament. There is no strong capital. Each German city has developed its own character and strengths. The former capital Berlin is a cultural and economic centre and Bonn is the present federal capital, but in summer 1991 the federal parliament voted to restore Berlin as capital. Hamburg, on the Baltic coast in the north, increasingly prone to flooding, is amongst the wealthiest cities in the Community and is known for shipping and commerce. Frankfurt is the financial and business capital with the headquarters of the main banks and the pre-eminent stock exchange. Other cities include Dusseldorf, Dortmund and Essen, the latter known for its heavy industry. The cities are linked by an excellent network of autobahns and railways.

Demography is a worry for the Germans with low birth rate and a much lower proportion of under 15 year olds (14.6 per cent in the west) than the Community average (18.8 per cent; UK 18.9). The proportion of over 64 year olds is the same as that of the UK at 15.5 per cent, which compares with a Community average of 14.0 per cent.

The federal structure

Within the framework of national guidelines, Germany's administrative structure follows the sound principle of handing the decision making process down to the lowest possible hierarchical level so that decision is taken closest to those affected. The country is organised into autonomous entities called *Länder* (plural of *land* pronounced 'lunt') which developed out of the old baronial states and retain their status as independent states. Each *Land* has its own economic development corporation with responsibility for advising, promoting and financing new industrial development and issuing its own construction regulations, the *Landesbauordnung*. In a similar way the *Länder* are themselves split into entities named *Gemeinden* which have considerable autonomy in such key areas as town planning, local facilities, and transport policy. Germany must be regarded as a series of genuinely independent regions – the *Länder*, which themselves have limited authority over the local *Gemeinden*.

Germany and the Community

As the most powerful Member State in the Community, Germany has always taken a leading role in Community affairs. Since reunification it has remained even more committed to European economic and monetary union, urging progress towards integration in Europe in the interests of stability. On the other hand, the sheer size and economic power of the country gives it a powerful voice, and the strength of the Deutschmark within the European Monetary System means that its economy and monetary policy, traditionally cautious and tight, affects every member in the Community. Germany and its policies and practices cannot be ignored by companies and individuals interested in a future in Europe.

The economy

The future of the German economy is uncertain. Either the massive investment in east Germany will bring rapid results and the stimulus will power the new state to greater strength, or the inflationary effects will lead to a drop in business confidence and a slowdown. At present the private sector is taking a cautious attitude to investment in the east, but many experts predict a surge of growth over the next five years. Certainly, the former Federal Republic of Germany economy appears to be enjoying dynamic growth. Optimism prevails in almost every sector. Orders and profits are strong, while favourable market prospects indicate a sound basis for continued economic growth.

Export and investment buoyancy add confidence for the future prospects.

Civil engineering has been less bright. A depressed market has hit companies operating in that field, mainly in relation to road and other public works. As a consequence the large companies which specialise in civil engineering have been diversifying and moving into other areas of the industry.

Housing related work is buoyant. People flooding in from the east presented the Federal government with severe housing shortages, in response to which it provided extra funds for the construction of social dwellings. New housing and rehabilition of old buildings have been in demand. Even old industrial buildings have been converted into flats. Extensions, adaptation of cellars and and attics, additional floors – all possible space has been utilised. The industrial building market has been overtaken by housing work.

The west German construction market is clearly attractive. It is the largest in Europe, with 12.2 per cent of GDP going into public expenditure. The following 1988 Euroconstruct data show how the west German market compares with that of the UK (Table 13.1).

The stability as well as the size of the German construction market is attractive. Germany has been able to maintain growth, despite, or because of, tight monetary policy operated by the independent Bundesbank, the central bank. Inflation has been successfully contained and the economy has been able to avoid the sharp fluctuations between growth and stagnation or decline which have characterised the British economy, with dire consequences for the construction industry. In investors' jargon Germany is a 'blue chip' market.

East Germany still has no 'market' as western economists understand it but plenty of work to be done. New building is picking up. The building stock is craving for maintenance and renovation. The natural environment is pleading for cleaning up and protection measures. The infrastructure is crumbling. The road and rail system needs invest-

Table 13.1 Comparison of construction output in Germany and the UK 1988

	Western Germany £ billion	UK £ billion
New building	33.3	21.5
Renovation and modernisation	27.7	15.5
Civil engineering	17.6	4.4

ment of at least ECU 100 million over the next 10 years, roughly equally split between the two. Some 55 per cent of the autobahns are in a 'catastophic condition', 68 per cent of east Germany's city streets and 22 per cent of the two-lane national road network need to be repaved, and 30 000 bridges need urgent attention.

Unfortunately, firms wanting to work in this corner of the EC find no job file on their shelves to refer to. Enterprise and courage are the qualities needed in this market. The preliminary financial, legal, and employment problems are daunting. German firms from the west will certainly be at an advantage, with neither language nor cultural barriers to overcome, though many east Germans are resentful of the wealthy western half of the country which has swallowed them up. The initial euphoria of unification has given way to resentment. Exploring the market early and making contacts will pay off; in the words of a headline to a *Financial Times* article about east Germany and eastern Europe, 'On your marks, get set, wait', but not for too long because others are getting in there.

The business environment

Disappointingly for the free-marketeers, an obviously successful Germany is not a good example of a free-market economy. The German economy is rigid, regulated and controlled. In Britain the subtext is 'everything is permitted unless it is expressly forbidden' while in Germany 'everything is forbidden unless it is expressly permitted'. Why is the entrepreneur and the innovator not stifled? In addition to the restrictive regulatory system, Germany is also hampered by large, conservative private organisations, particularly the banks which dominate the business scene. At a national or local level they are the main source of funds, but unlike the British financial institutions which chop and change between equity holdings looking for the short-term advantage, the German banks tend to think and lend long-term. Government and family holdings also provide stability. There is little opportunity for speculation in company shares, and with the stock markets playing only an assisting role, the economy remains monotonously strong.

There are compensations. While British management may spend time staving off hostile takeover bids, their German counterparts devote themselves to technical excellence and thorough management. They can afford to take a long-term view as there is less pressure for immediate return on investment. Wheeler-dealing is not part of the German culture. Managers prefer to concentrate on achieving long-term growth internally rather than through acquisitions or manipulation of assets.

In addition, a partnership between the Government and industry exists, with close co-operation contributing to maintenance of economic stability. This symbiosis is preserved by an innate distaste for stepping out of line. Those who break rank find themselves quickly isolated.

The construction industry

As in most countries the industry is structured along familiar lines, with few large and many small companies. Only the largest contractors operate outside their own *Land* or even locality, with most construction activity being made up of local business. The construction industry remains essentially local and traditional. Table 13.2 lists the five largest German contractors and their 1987 output.

Ownership patterns in Germany are complex, often involving banks and other financial institutions, other companies and government. The industry is considered a safe investment. This makes it virtually impossible for a foreign firm to acquire a stake in a German company. Instead the most hopeful strategy for penetrating the market is to establish some kind of collaboration with a local partner which can provide expertise on local regulation, planning, and labour.

The regionally segmented market and the rigidity of the funding system have preserved the position of the small and medium-sized family owned business. These companies often retain the paternalistic approach and suffer from the lack of a multi-disciplined professional management. Outside investors have to take great care in undertaking any acquisition as the owner may be concerned about betraying his workforce. One British businessman found himself negotiating a deal

Table 13.2 Output of the major German construction companies in 1987

Company	Domestic	Foreign ECU million	Total
Philipp Holzmann AG	2019.9	1562.9	3582.8
Hochtief AG	1577.8	652.0	2229.8
Bilfinger & Berger AG	773.4	744.1	1517.5
Strabag Bau AG	805.0	152.9	957.9
Dyckerhoff & Widmann AG	767.9	142.1	910.0

in a forest because the owner wanted to ensure that he was not suspected of selling the company. The long-term welfare of the employees and the social responsibility of the firm is of great concern.

The German trade unions have an excellent reputation for responsibility and collective bargaining. Any firm employing more than five operatives must compulsorily set up a works council. The ability of the unions to maintain a constructive relationship between management and employees is the key to retaining a reliable and permanent workforce. The high levels of education and training, as well as the respect given to formal qualifications, also help in establishing mutual respect. There is a much less marked social chasm between owners, management and operatives than seen in the UK. Surprisingly, German safety standards are lower than those in the UK (the fragmented system of safety regulations and enforcement which exists in Germany is described in chapter 7).

The system of training in Germany is highly organised and developed. It results in operatives who are among the most skilled in Europe, though the emphasis on training and qualifications at all levels can result in lack of flexibility. Manual and technical work are regarded on a par with intellectual and scientific work. In Germany, the two pillars of a civilised society are *Kunst* (art or culture) and *Technik* (a term which does not have a direct translation in English, but can be loosely be described as the science of making things) and they are treated with equal respect. The British cultural divide does not exist. Good practical work is highly valued. The high levels of respect accorded to engineers is another symptom of the same phenomenon and explains why the proportion of graduates in engineering is so high.

The German construction industry produces a very high quality of output. This is particularly true in the mechanical and electrical engineering sectors. However, in consequence goods are more costly. In addition, the emphasis on quality work results in building which is slower than in other Member States. German companies find it hard to compete outside their domestic market. But at home clients are prepared to pay for top-quality materials and top-quality workmanship.

Cost control is less advanced than in the UK, and cost overruns of up to 20 per cent of the project budget are not uncommon. In this area, British quantity surveyors may be able to sell their services to German clients. However, crude cost per square metre data would not impress German clients, and the quantity surveyor would have to adapt to the demand for the highest quality and standards of technical excellence.

French and German management styles present an interesting contrast. Germans instinctively respect laws and rules. They will always opt for a standard solution if possible. German managers feel uneasy when faced with uncertainty, ambiguity and the unknown. Cutting corners is frowned on. Opportunism is regarded as demonstrating a failure to plan and organise, whereas the French respect it as a sign of adaptability. People from more laid-back cultures must be cautious about making commitments – in Germany, one's word is always adhered to and commitments carried out to the letter. Business dress shows more variety than in the UK, but manners and address are formal, with first names rarely used. The Germans take their humour seriously, and distrust the mixing of humour and business. This attitude is also exemplified in advertisements. British advertisements always strive for humour, while Germans would not be impressed – the message is generally facts and quality.

Although the level of education amongst the business and professional community in Germany is perhaps the highest in the Europe, the Germans rarely create the multinational companies often found in the Anglo-Saxon world, and the strong export industry is instead largely built on home-based manufacturing.

High quality work carried out by competent staff characterises Germany. Strict prevention of building defects is the German way of keeping the risk of problems after completion to a minimum. Thus the builder's liability period following acceptance of the work by the client is shorter than anywhere else in the Community. The warranty period for buildings is either two or five years from acceptance, while for infrastructure works it is only one year. All parties in the construction team are expected to be truly competent. Competence and the avoidance of fault is regarded as more important that establishing liability.

Contracts

Generally, clients enter into separate contracts with each of the specialist contractors. Work is also commonly carried out by two or more contractors in joint ventures, even on relatively modest contracts. This provides greater assurance for the client and makes more resources available. General contracts, management contracting and turnkey contracts remain exceptional. Lump sum contracts are only used when the work has been defined exactly in the contract documents and no variations are expected, and when the work is of very limited duration. Property development as found in the UK is almost unheard of. Contractors rarely produce designs, but expect to work to detailed working drawings and technical specifications.

Construction contracts in Germany have a uniform structure. General conditions prepared by the German Committee for Tendering and Contractual Procedures for construction work and developed over the last 50 years are used. They are compulsory for government works and widely used by private clients. These contract conditions are referred to as *Verdingungsordnung für Bauleistungen* or simply VOB. In these, the general provisions of the Civil Code are replaced by the particular provisions of the VOB. The VOB conditions are often implied into contracts even when not specifically agreed.

Architects and engineers

The titles architect and engineer are protected by law, and fee scales are regulated by 'ordinances'. Price competition is thus eliminated, and competition is based on experience and resources. The various tasks performed by the parties are strictly fixed. It is not the time spent on a project or the quality of the drawing paper which matters, it is the end result: the object built.

Standards and approvals

The German system may appear protectionist, but it is inspired by liberal principles. Statutary requirements and regulations are onerous, but are in fact less restrictive that the multiplicity of obscure quasi-regulations found in certain other countries.

Contractors, project designers, all parties involved in the construction process must comply with the DIN standards (Deutsches Institut für Normung) which are prepared by mixed industry-science-government committees. Deviations from these are only permitted if safety can be proved. DIN standards represent the toughest standards in the Community (in relation to building this is partly because of the relatively severe German winters). The DIN standards are regarded by many as a point of reference, and because of this are used by some firms which operate in a number of Community countries for all their work. Newcomers should recognise that performance requirements for structural stability, thermal and sound insulation and weather tightness are particularly rigorous. Standards in concrete construction are also high. Much of the early research into reinforced concrete technology was sponsored by the German cement industry. Undoubtedly, the tough German quality requirements represent a challenge for foreign companies seeking to establish themselves in the market.

Liability

Contractors bear full liability since insurance against the consequences of mistakes is not available. Professionals also remain responsible for their errors. The parties are thus anxious to ensure that the work is done correctly, as reflected in the quality of German buildings.

In 1985 a new standard contract for architects was introduced which provided for a 15 year liability period starting on acceptance of the building by the client. However, this period may by agreement be reduced. Architects in Germany have to ensure that accepted good practice is observed, including that relating to thermal insulation and to external and internal acoustic insulation.

Contractors are obliged to correct defects which appear during the guarantee period, but only if the client orders them in writing to do so.

Disputes

Disputes are rare. Builders tend to meet their client's demands and carry out repairs free of charge beyond the contractual liability period. Settlement of disputes through the courts is always laborious and a conclusion may take many years. In any case, builders wish to avoid the adverse publicity that disputes bring, especially in the regions where clients and contractors know each other. Out-of-court settlements usually reflect the relative strengths of the parties rather than the facts of the dispute.

Conclusion

The German construction market is highly regulated. Standards are stringent and only first-class workmanship is acceptable. Procedures are established and regulated by government. The industry is fragmented regionally. The system impresses by its coherence and Germans are confident that the system is the best in the Community.

Germany is attractive for individuals wanting to find employment in the construction industry. Conditions are good and the salary levels high. The statutory working week is 48 hours though in practice usually less. There are 10 to 14 statutory public holidays, depending on the region, while five to six weeks holiday are normal, much in excess of the statutory 18 days.

FRANCE

France is the largest country in the Community and with its 55.9 million inhabitants has a density of population less than half that of the UK. Its central position in the Community gives it a special advantage.

France is a country of strong traditions and cultures. The inhabitants of Alsace, Brittany, the Basque region and Provence vary markedly in attitude and character. They have in common a love of their regions, to which they are so attached that relocation is difficult, even when work requires. A posting to Paris is almost felt to be expatriation and French managers secretly envy the mobility of the British.

Like London, Paris is the political and business centre of the country: 64 per cent of all French companies have their headquarters there. Other important cities are Marseilles, Lyon, Lille, Bordeaux, Nice, Toulouse, Grenoble and Strasbourg.

France and the Community

Although intensely patriotic and proud of their country, language and national achievements, the French were the prime instigators of the EC. France has produced notable figures such as Robert Schuman, initiator of the European Coal and Steel Community in the 1950s. Currently, Jacques Delors, President of the Commission, is a driving force behind '1992', and is striving towards monetary and political union. The French, though strongly pro-European and believing in a federal Europe, are renowned for protectionist tendencies and a powerful, often militant, farming community.

As a result of French prominence in the Community, there is in France a tremendous awareness of the approach of '1992'. Listening to the media one soon realises that France is on red alert. Magazines and newspapers hotly debate the implications of the single market and its impact on every aspect of life. *Le Grand Marché* is a fashionable topic and attracts interest. Radio stations broadcast advertising campaigns promoting foreign language learning packages. France has the best record in implementing EC directives, demonstrating its commitment to the Community and 1992.

Administrative context

France is a country with a written law and written constitution. Centralism prevails as a political response to France's large size and diversity. The basis of this centralism is a rigidly codified legal system: the *code Napoléon*. Codes are common approaches to organisational problems, but in France are not as closely followed and respected as in Germany. The code for government contracts, the *code des marchés publics*, a document of 377 articles, is a good example of French codification. The country is still run on dirigist lines. Through a series of national plans the government plans and guides economic and social development. Industry diligently follows.

The country is split into 100 *départements* which are grouped into 22 regions. There are some 36 000 *communes* which often act on behalf of the state. Unlike in Germany, the regions and the departments have no say in law-making. At central level, the *Ministère de l'Equipement* (the equivalent of a combined Department of the Environment and Transport) remains powerful despite a decentralization process initiated by the state. At regional level, the *Directions Départementales de l'Equipement* or *communes* for large towns substitute for the state.

The economy

Like other EC countries, France is struggling with inflation and a balance of payments problem. High unemployment threatens social stability and it is proving difficult to maintain the value of the franc against the all powerful deutschmark. However, as a whole the economic outlook is surprisingly attractive for business. Buoyant exports and strong industrial investment sustains the economy and bodes well for the future.

Table 13.3 Comparison of construction output in France and the UK 1987

	France £ billion	UK £ billion
Housing	9.68	6.75
Repair	16.62	15.54
Commercial	6.01	8.44
Public	11.93	3.88

Source: Euroconstruct

France strikes one initially as a socialist country, the British might say frighteningly socialist. Vast amounts of money are injected into public services: some 18 per cent of GDP is spent on public procurement procedure. Incidentally, protectionism still exists despite the EC public procurement directives. In terms of construction, France is an attractive market (see Table 13.3).

The house building sector remains steady. Renovation and modernisation work shows a relatively strong growth with the residential renovation market showing the way. New office construction and industrial investment expenditure are expected to level off slowly. Civil engineering work in the public sector seems to be growing. Much activity is sustained through large construction sites such as the Channel Tunnel and Eurodisneyland and building near railway stations. The highly successful TGV (*train à grande vitesse*) network is being extended and integrated into part of a European network. Major projects involving road and airport construction, canals, new electricity, water and cable networks will increase activity. Generally speaking, optimism in most construction sectors is high.

The construction industry

France already has a well developed construction industry which can compete directly with that of the UK. A large number of very small firms carry out local work, while a number of extremely large firms such as the giant Bouygues carry out the larger projects. Table 13.4 lists the six biggest French contractors and their 1987 output. Large firms are particularly significant, the medium-sized sector is relatively weak and the situation regarding small firms is similar to that in Germany.

Sub-contracting is less common than in the UK. Many projects are dealt with *en lots séparés* (through trade contractors) with the co-ordinating task left to a *maître d'oeuvre* who may be an architect or an engineer. Unlike the UK, there are few house builders in France since most houses are built using independent trade contracts.

Large firms usually specialise in working as main contractors (known as *entreprises générales*) as in the UK, although they may have specialised subsidiaries. These construction groups have benefited from extensive experience in many parts of the world. Some French companies have close relationships with Spanish and Italian counterparts who tend to work in a similar way.

In France companies evolve in a highly competitive market where few foreign firms are active as the result of protectionism.

These large organisations tend to have complex holdings in each other which makes acquisition in France difficult. Ownership patterns

Table 13.4 Output of the major French construction companies in 1987

Company	Domestic	Foreign ECU million	Total
Bouygues	5764	1538	7302
Société Auxiliaire d'Entreprises	2233	1365	3598
SGE Groupe	2363	616	2979
Spie Batignolle	2094	821	2915
Dumez	1348	1393	2741
GTM Entrepose	2185	245	2430

are often incredibly complex and direct comparison with UK contractors is more than troublesome.

Large French companies are masters of strategic planning. This is based on logical induction, starting from a vision of the future, supported by expert analysis and detailed economic study. The plans are elaborate, far-reaching and detailed.

Banks are often major shareholders of large construction groups. This state of affairs provides a financial stability for the firm and enables the implementation of the long-term corporate plan. Operations such as moving into another Member State's market tend to be meticulously planned through carefully negotiated acquisitions, mergers and joint ventures. However, because of this strong commitment to planning, French companies may find themselves utterly out of tune with foreign business surroundings. The development of markets does not necessarily follow a rigid logic. It proves that there are pros and cons to long-term planning.

On construction sites, things are completely different. French engineers have a reputation for flexibility and improvisation skills. It is a balance between pure professionalism and what is commonly known as the system 'D' (d for *se débrouiller*, the French for 'to manage'). Building operations are less planned than in the UK in both managerial and financial terms. (It would seem that British quantity surveyors have much to teach French construction economists.) Should an unforeseen event take place, a decision is likely to be taken fairly rapidly without referring scrupulously to the conditions of contract or a resident engineer. Less attention is paid to the content of the contract than it would be in the UK. Achieving a good result matters more than merely fulfilling a duty.

In a pure Latin tradition, many quarrels between architect, engineer, and client are resolved around a table, usually at lunch time just before the weekly site meeting. What takes place in such circumstances might be regarded by the British as corruption.

Building regulations

In France, what is compulsory and what is not is often a mystery. There are no clear and simple 'building regulations' but dozens of obscure and lengthy rules. Various types of texts ranging from the statutory to quasi-statutory are applicable.

Private inspectors

In France, as in Belgium, there are private inspectors who to some degree counterbalance the lack of governmental control over building. The cost of this extra party is of the order of 0.5 per cent to 1 per cent of construction costs. Private inspectors have to be approved by a committee set up by the state on the basis of their competence and their professional ethics. Technical inspectors are called in when their service is compulsory, as it is for certain forms of construction which present particular hazards to the public, when the insurance premium price is reduced by their involvement, or simply when a client chooses. The presence of technical inspectors tends to reduce the authority of the project manager and may delay site decision making.

Contractor sub-contractor relationship

The builder or the main contractor remains responsible to the client regardless of the appointment of sub-contractors. Relationships between principal contractors and sub-contractors are always difficult to grasp. The sub-contractor provides a warranty under his contract with the principal contractor and voluntarily takes out liability insurance. Contracts are often not 'back to back' as is the case with the JCT Standard Form of Building Contract used in the UK.

Designers

A project designer is responsible for the architectural design which must accompany the application for a construction permit. His responsibility depends on the extent of the task which has been entrusted to him. This could be a simple design task, or cover design and supervision of the work.

There are different standard contracts between architects or *bureaux d'études techniques* (design office) and clients. Relationships are frequently based on the *décret* of 28 February 1973 which specifies tasks and remuneration for designers. They are referred to as tasks m1, m2, m3 for supervision and tasks m6, m7, m8 for design.

Liability

France, like Belgium, has adopted a 10 year liability – the *responsabilité décénnale* – which is guaranteed by an insurance policy. This is compulsory for the client and for every builder. This particular approach provokes much controversy in European building circles since the French are keen to harmonise the construction market and are trying to get this system adopted throughout the Community. The British professionals claim that the system discourages innovation, since the insurers exercise a heavy hand on the industry and on design. Incidentally, in order to obtain insurance from a French insurer it is necessary to obtain approval from one of the two private agencies which cover the construction sector. These are the Office Public de Qualification pour la construction des bâtiments (OPQCB) for building contractors, and Service de Recensement des Entreprises de travaux publics (SRTP), which is part of the Fédération Nationale des travaux publics for civil works. In both cases approval takes about three to six months to obtain.

The *décennale* system originated under the pro-consumer era in France of the 1970s, mainly from exasperation that so many construction problems were not being rectified because of the insolvency or winding-up of the builder. Under the system insurance is compulsory for new construction, renovation works and the restoration of old structures. All items which are integrated into the structure, such as the foundations or roof, are subject to the 10 year liability and compulsory insurance.

This obligation does not exist for what is called in France a *constructeur de maisons individuelles* (house builder). There is no direct protection for a 'small client'; he is expected to take out his own insurance policy.

Otherwise, there is a warranty of satisfactory performance (known as the *garantie de bon fonctionnement*) spanning two years. It applies to the 'dissociable' elements of the construction to the exclusion of domestic equipment and other machinery. Lastly there is the warranty of completion (*garantie de parfait achèvement*) lasting one year. The three warranties begin from the date of acceptance. There is no requirement to insure in respect of the *garantie de bon fonctionnement* and the *garantie de parfait achèvement*.

The system in practice

The principle of the insurance system is to provide insurance against damage for which the builders are legally liable for 10 years. In practice, when damages occur, repair works are carried out rapidly without determining liability. Responsibility is provisionally shared and settlement is made afterwards. The insurance costs are often paid in a single lump sum before work begins on site or as an annual premium. The cost of insurance is of the order of 2.5 per cent of construction costs, exclusive of the cost of the technical inspector.

Conclusion

Good knowledge of French is indispensable to operate in France. One needs to be skilled on a higher level than *une bière, s'il vous plaît*. Technical terms and a plethora of abbreviations used in a typical French contract must be mastered.

Working conditions in France are very appealing for employees. Salary wise the *Daily Telegraph Guide to Working and Living Overseas* reckons that the level of managerial salaries is at least 40 to 50 per cent higher than in the UK. Employer's social security contributions are the highest in Europe while direct taxation is about the lowest. There is a statutory regulation which imposes a working time limit of 39 hours. Naturally, long hours and hard work are expected in managerial positions, though workaholism is not particularly valued. French people believe that there is more to life than the job. Weekends and vacation days, sport, cultural activities and family life are very important. There are 11 public holidays and 30 days statutory paid annual leave. In construction companies, the decision makers are very often *ingénieurs* of the *grandes écoles*. Engineers have power and high social status. On a more general level, colleagues are expected to be lively and interesting companions, well informed and appreciative of the good things in life as well as successful at their job. A laid-back attitude to life coupled with professionalism are present at the work place.

ITALY

Italy has a population of 57 million and land area 25 per cent larger than that of the UK. There are two officially distinct regions. The wealthy north is centred on Milan, the financial and industrial centre of the country, which is acknowledged to be like a 'big village' in terms of business relationships. Milan, Turin and Genoa contain about half Italy's population. The people of the north are said to be preoccupied with work and money. The south, the *Mezzogiorno*, with 19 million people, begins just south of Rome and is economically underdeveloped. The main cities in the south are Bari, Naples and Taranto. Northerners like to include Rome in the *Mezzogiorno* and tend to consider the southerners as culturally closer to Africa than to Europe. The economic despair of the south is reflected in the average unemployment rate: 21 per cent in the South as opposed to 8 per cent in the North.

Italians (known to the Germans as *Spaghettifresser* – spaghetti stuffers) have over the last 10 years become true believers in a federal Europe, yet, apparently paradoxically, their government has had a poor record in implementing directives. A common explanation is that Italians have a poor view of the efficiency and effectiveness of their own state and think that a federal Europe could only do the job better.

Although the country has a reputation for creative flexibility, bureaucracy can be impossibly frustrating for foreigners wanting to set up locally. The red tape and inefficiency are renowned.

For Italians in business personal contact is the crucial factor; trust and reputation are fundamental. In commercial dealings it is the individual rather than the company which counts. A unique form of specialist intermediary found in Italy, the *commercialista*, a mix between a financial and legal adviser, can be invaluable when trying to arrange deals and seek administrative shortcuts.

The economy

The most notable feature of the Italian economy is the regional disparity between the north (and centre) and the south. The north has a purchasing power parity above France, Denmark and Germany, while the south is similar to that of Spain and not much above that of Greece and Portugal. The disparity has been present since unification in 1860 and has remained despite government attempts to improve economic conditions in the south. Output per employee in the south is only 56 per cent that of the north. The disparity is particularly marked in unemployment. In the north the overall level was 7.7 per cent in 1988 while in the south it was 20.6 per cent. Among 16 to 24 year olds the disparity in unemployment was even greater, at 23.3 per cent in the north-centre and a staggering 55.3 per cent in the south. Unemployment in this age group had been steadily rising between 1982 and 1988.

Overall economic growth in Italy has remained healthy, averaging 3 per cent per annum in the five years to 1988, 3.2 per cent in 1989, 2.6 per cent in 1990, and OECD forecasting 2.4 per cent and 2.7 per cent for 1991 and 1992 respectively. Overall unemployment is marginally above 11 per cent and likely to stay there. Inflation is running at a lower level than in the UK, but still above the Community average. Manufacturing output and exports remain strong. Construction output has grown slowly, reaching a peak of 3.8 per cent in 1988, except in housing which fell slightly. Output in 1989 was 3.6 per cent up on 1988, but was expected to slow, and even become negative in some areas in the period up to 1992. This is due to the general slowdown in economic activity but particularly due to expected falls in public sector orders since the government is having to restrain public spending.

The construction industry

Employment in the construction industry at 9.4 per cent of total employment is larger than the community average, the percentage in the south being rather higher than in other parts of the country. In 1984, 6.4 per cent of employment in the north was in construction,

Table 13.5 Output of the major Italian construction companies in 1987

Company	Domestic	Foreign ECU million	Total
Lodigiani SpA	588	84	672
Halstrade SpA	399	42	441
Cogefar Construzioni Generali SpA	196	110	306
Recchi SpA	16	282	298
Impregilo SpA	0	244	244
Girola SpA	192	48	240
Astaldi SpA	94	140	234

while in the south it was almost twice as high at 11 per cent. The industry is highly fragmented, comprising a handful of large groups and thousands of small to medium-sized firms, mostly privately owned family businesses. It is estimated that there are more than a quarter of a million Italian contractors, of which 40 000 can bid for public sector contracts. Italian builders are internationaly esteemed. Italian building material producers enjoy a worldwide reputation, especially for marble and ceramic products.

The vast majority of contractors, including some of the largest companies, are family owned or controlled. These families jealously guard control of the management and are very unlikely to sell or purchase stakes in rival companies unless able to gain managerial control. Foreign contractors therefore find it difficult to gain a foothold in the Italian market through acquisition of strategic stakes in local companies. Table 13.5 lists the seven largest Italian contractors and their 1987 output.

Tendering for work

Italian tendering procedures for public works (see chapter 6) are often described as chaotic. The system is fragmented, highly inefficient and closely linked to politics. At times the Mafia may also be involved, though this is not generally true in the north. Contracts tend to be placed locally, with little regard to price or ability to complete on time. There are more than 14 000 local and state organisations who have the authority to undertake capital works in construction.

Conclusion

European construction related companies will find it hard to break into the Italian market, whose fragmented nature and heavy reliance on personal contacts makes it difficult for outsiders to penetrate. Most Italian contractors believe that the European single market will have little impact on established patterns in Italy. They are certain that they will continue to win most of the available contracts.

UNITED KINGDOM

The United Kingdom is a constitutional monarchy of 57 million inhabitants, is divided into England, Wales, Scotland and the province of Northern Ireland. Some 82 per cent of the population is concentrated in the south of England. The southeast region has a population of 17.3 million and a land area less than Belgium, giving it a population density of 636 per square kilometre. Holland, the most densely populated country in the Community, has a population density of 356 per square kilometre. On the other hand, Scotland, with a population of 65 per square kilometre, contains some of the few remaining wildernesses in the Community.

Administrative structure

The administrative structure is, like France, heavily centralised in the capital. The political system is based on 53 counties and 370 districts which have no great power. Unlike most other European countries, the United Kingdom relies on an unwritten constitution and a customary law system. In the UK, thus, there is neither a bill of rights nor civil code. However, there is what is called 'common law' which is a legal system based on previous decisions taken by the judiciary. Particular requirements considered necessary for society are expressed through 'statute law'. Such laws are referred to as Acts of Parliament.

The UK and the Community

Generally speaking, Britain is not known as a committed pro-European Member State and this does reflect in the Continental media. The British government, rightly or wrongly, is singled out for not playing a constructive role in the Community. Attitudes to the Channel Tunnel illustrate the problem. The idea of linking the Continent to the island of Great Britain by a fixed passage brings a deep sense of pride to the French as well as hope and enthusiasm for the future. The British reaction tends to be cooler. The tunnel will significantly shorten journeys from London to Paris and London to Brussels to respectively 2 hours 40 minutes and 2 hours 30 minutes and prevent the Continent from being cut off by bad weather! but suspicion prevails that some terrible disease will be introduced and that national sovereignty will escape through it.

The British are characterised by early development of nationhood. As in France, the concept of the 'nation' is intimately fused with that of the 'state'. The British people appear to be unable to reconcile national and European identity. In contrast, Germany, Italy and Belgium, for whom nationhood has never been as deep, can cede sovereignty to the European Community without much anguish.

Michael Toner, in his humorous *Bluff Your Way in the EEC*, argues that the British are still traumatised by the loss of empire and only joined the Community because they believed they could run it instead. To the concern of some, things have not turned out as expected and it has come as a surprise that colonial attitudes have found no positive response. Has the colonial attitude disappeared? Gabriela Iglesias RIBA in a seminar 'Working In Europe' (October 1990), reproaches some British professionals working in the Iberian Peninsula for colonial attitudes which impair business relationships. The failure of many British to be able to communicate in other languages is perhaps indicative.

The British

In Europe the British are considered to have the best sense of humour. The 'Rosbifs' to use a current nickname given by the French (literally the roastbeefs) are regarded as sometimes gauche with relationships amongst themselves and certainly with others. Representatives of French companies dealing with British counterparts experience the frustration of fundamental misunderstanding. British values such as gentlemanly amateurism, common sense, politeness based on the avoidance of confrontation, and team work, often contrast with the features of citizens of other Member States such as analytical professionalism, verbal and physical expressiveness, and individualism at work.

Business climate

The United Kingdom offers an attractive environment for industrial investors. The UK market has become a gateway for many non-EC firms into the European single market. A third of all Japanese manufacturing facilities in Europe are located in the UK. However, since foreign firms tend to feel most comfortable working with construction firms with whom they have enjoyed longstanding relationships, many of those investing in Britain may choose to ignore local construction capability.

Britain has much to offer. Labour costs are generally lower than other major economies in the Community, with wages and benefits well below the norms in Germany and France. Free market ideologies predominate and there is little trace of antagonism towards outsiders.

There is little or no regulation concerning working time, overtime maximum, public holidays and paid annual leave. Redundancies are not difficult to implement when they have to be made.

The economy

The UK economy seems to be characterised by more pronounced cycles than other European economies. Only two years after the boom year of 1987 the word recession is on everybody's lips. Business confidence has collapsed. Interest rates have caused consumer spending and industrial investment to fall. Only firms exporting to thriving markets such as France and Germany prosper and remain optimistic.

Overall construction activity in the UK is on the decline and rather more sharply than the general downturn in the economy. In 1987 and 1988 the industry could not find employees to cope with the workload, yet at the start of 1991 the talk was of 100 000 jobs to be lost in the industry. Company failures and bad debts have increased as interest rates have reduced demand for residential and commercial properties. Stocks of unsold bricks have risen to a disturbingly high level. Job Centres have been kept busy as company failures and redundancies have mounted across the construction sector. The UK depression is driving firms related to the construction industry to seek work outside the national boundaries to survive. The European single market is the most obvious place to try, particularly as the USA also has serious economic problems.

The outlook for the industry is not entirely bleak. From the early 1990s, the forecast is for a boost to civil engineering activity with many major infrastructure projects in the pipeline. Some of these will be privately financed, for example in the transport sector and the newly privatised water industry.

The construction industry

With France, Britain has the most developed construction industry in Europe. In 1988, the six largest EC contractors ranked by turnover were British or French. A large number of small firms carry out local work, whereas a number of powerful companies deal with larger projects. In short, large firms are particularly significant; the medium-sized sector is relatively weak.

The structure of the construction industry reflects the conditions of contract it operates with. The British tradition is for construction work to be carried out by a main contractor and a fleet of sub-contractors, some of whom, termed 'nominated sub-contractors', are employed by the contractor on the instructions of the client's representative. The widespread use of main contractors contrasts with France and Germany where a project is likely to be split into separate lots.

The extensive use of sub-contracting has left the small to medium-sized companies almost unable to perform 'solo'. In these set-ups a whole layer of management is absent and that layer is to be found within the supervising organisation, namely the main contractor. Consequently it seems that a French or German specialist firm may encounter fewer problems operating in the UK market than British counterparts will on the Continent.

British firms generally think short-term and would not contemplate establishing a strategic plan. Their financial structure tends to mean that immediate results are required. A strong desire to make quick profits combined with a lack of long-term commitments are apparent in the accounts of most British firms. For a start, accountants within an organization have a prime role in running the business, whereas in France and Germany technocrats rule. A company division or subsidiary may be severely cut back or disposed of on purely financial grounds and with apparently little reflection. High return on the initial investment is paramount. When EC contractors are ranked by pre-tax profit, the first 10 are British.

This characteristic can conversely be regarded as conferring certain strategic advantages. British contractors, for instance, appear to adjust more easily to cycles in the construction market than their Continental counterparts who are burdened by centralist planning systems. Construction companies are adept at moving from one specialism to another, for example from traditional contracting into commercial and industrial development, or housing. Sometimes, however, the new activities can bring their own problems to those who rush in. In 1973 and 1990 many construction companies operating in property development and housing ran into difficulties when market conditions changed sharply. In the UK, pragmatism and flexibility prevail over long-term logic and intricate planning.

Contract

Those wanting to break into the UK construction market must be aware that the conditions of contract are a crucial element in the process of trading. The inherent tendency, referred to above, to avoid verbal confrontation is counteracted by the use of sophisticated contracts elaborately defining roles (in which the law has almost nothing to say). In the event of disagreement the conditions will be scrutinised by highly paid members of the legal profession and will be applied to the letter by the judiciary. In fact, it is even more complicated than that – the legal values of a contract go well beyond the words in the document itself. Past cases are referred to in disputes, serving to provide both additional meaning to the conditions and to undermine seriously the clarity of the conditions. Since the legal aspects of the contract weigh heavily, meetings relating to the contract may be crowded with various levels and types of technical and legal representatives. (In the UK there are around 59 000 'registered lawyers' (barristers and solicitors) in comparision with around 17 000 in France.)

The principles of contracts in the UK differ fundamentally from those in France. There is no *code des marchés publics* to define rigidly contractual relationships, a minimum of limitation is provided by the statute law. A prime example is the Unfair Contract Terms Act 1977. This has the effect merely of rendering void clauses which seek to exclude certain types of business liability. In France, in contrast, a construction contract which does not conform with the legal requirements may be considered as void. Contractual matters and related liabilities also hinge on other statute law such as the Defective Premises Act 1972, the Sale of Goods Act 1979 and the Latent Damage Act 1986.

Further vagueness is introduced by the diversity of procedures such as the traditional approach, design and build, develop and construct, turnkey, management contract, project management, client managed work, and standard forms of contract produced by the Joint Contracts Tribunal (for building), the Property Services Agency (part of the Department of the Environment), the Association of Consulting Architects, the Institution of Civil Engineers, the British Property Federation and others. This multiplicity of systems is compounded by freedom to amend forms of contract or to produce new 'one-off' forms. Communication, team work and common understanding between the many parties is very frequently absent. This fragmentation and isolation tends to lead to major disputes as well as a high incidence of defects and delays.

Main bodies

Most central government construction work is handled by a part of the Department of the Environment – the Property Services Agency (its monopoly has recently been overthrown by allowing government departments to choose private sector organisations to act on their behalf). The Department of the Environment deals with planning and construction, and has in the past been responsible for the construction and maintenance of central government buildings and works such as Ministry of Defence properties and law courts. This influential organisation prepares draft legislation and regulations.

Other bodies

Unlike the German state, British government intervenes little in professional matters. It follows the tradition that anyone, unless clearly prohibited which is rare, may exercise any profession of his choice. The state does not regulate the activity of designers, entrepreneurs or tradesmen. However, this is counter-balanced by the principle in British law that those who undertake work are liable for damage caused by failure to exercise the level of skill and care of those who normally undertake that type of work. Damages for defects and consequential damage may be substantial, and most professionals maintain costly insurance policies offering professional indemnity. There is in general, though, nothing to prevent a professional choosing to 'go bare', an American expression meaning that no insurance is held on the grounds that without insurance one would not be worth suing.

In Britain the word 'professional' in this context refers to a party involved in construction but who only provides intellectual services. Engineers, surveyors and other professionals are normally members of organisations, which are not government regulated, but which act as qualifying bodies, setting standards of education and training for entry to the body, possible requirements for 'continuing professional development' and standards of professional conduct. These bodies also act as learned societies, promoting and disseminating research and knowledge into the activities of the profession. Leading examples include:

- The Institution of Civil Engineers;
- The Institution of Structural Engineers;
- The Royal Institute of British Architects;
- The Royal Institution of Chartered Surveyors.

Apart from the title of 'architect', which is protected by law, there are no legal requirements relating to other titles such as 'engineer' or 'surveyor'. Compared to those in most other Member States, British architects in building and engineers in civil engineering have considerable power concerning the acceptance of work. British professionals enjoy a reputation throughout the world of providing a relatively high standard of service, but are often seen as hindered by insensitivity to local cultural values and working practices abroad.

Quantity surveyors are an unknown quantity in mainland Europe. In the British tradition these are member of the 'design team' whose tasks include production of the 'bill of quantities', which is generally one of the contract documents, and the provision of advice on cost implications of design decisions, budgeting and the assessment of claims by the contractor for additional money, a characteristic part of the British construction scene. The quantity surveyor is usually directly employed by the client. On the Continent all the professionals have a clear function in relation to the completed structure, whereas the quantity surveyor has a role only in relation to the contract and appears to be an unjustified overhead. Many believe that the quantity surveyor will have to fight and adapt to survive in European construction.

In the UK few have a house constructed for themselves in separate lots. House construction is almost exclusively carried out by developers or house-builders who build for no particular purchaser. Some 27 000 house-builders are registered with the National House Building Council. In addition there are about 3000 developers. Most houses are constructed by builders registered with the National House Building Council, a trade body which carries out site inspections and organises a 10-year warranty covering structural defects.

For larger building projects, one of a variety of contractual arrangements may be used:

- The traditional approach where a general contractor tenders on the basis of restricted competition for work which is designed entirely by others. This method is still predominant.
- The management contract which made substantial inroads into the market, particularly the commercial and industrial sectors, in the 1980s. In this method the contractor is part consultant to the design team and part building contractor, except that almost all the work on site is carried out by trade sub-contractors for whom the contractor gets reimbursed at cost.
- Design and build, which is experiencing a considerable growth.
- 'Construction management' which is relatively new in the UK but very similar to much Continental practice. In the UK the construc-

tion management service is generally provided by separate consultants. Various US firms are particularly prominent as construction managers.

Liabilities

The situation on liability in the UK is particularly confusing and confused. Liability can derive from both the contract, under which only other parties to the contract may recover damages, and tort. Tort, known in Scotland as 'delict', is literally a civil wrong, but has become a branch of law mainly developed as a system of continuously changing rules derived from judgments in previous cases. This system of using law developed by judges is known as common law. Even contracts cannot be interpreted without reference to the continuously growing body of case law.

The law concerning liability, and the relationship between liability in tort and in contract has undergone at least one volte-face in recent years, with disastrous consequences for some. Liability under contract may extend to six or 12 years from the date when the particular contractual obligation was breached, depending on the technicalities of the way in which the agreement was prepared. In tort, the type of damages recoverable has generally been restricted, and in general liability may, under the Latent Damage Act 1986, extend to six years from the date when damage first occurs or three years from reasonable discoverability, whichever is longer, subject to a maximum of 15 years from the date of the breach. In comparison to most other northern European countries, liability for construction related work is complex, fluid and appears to be against the interests of the clients of the industry and others who suffer loss.

Conclusion

The UK system appears disordered and variable. External regulation is at a minimum and the contract documents are the primary base of communication and the business relationship. Uncertainty is inherent and the largely unwritten law is volatile and difficult to grasp. A no-nonsense pragmatism characterises the construction industry. Litigation is frequent but obtaining a satisfactory remedy can be a slow and costly business with almost no certainty of the outcome.

Those wanting to become employed in the UK will be surprised by the conditions. Executives and engineers are poorly paid by European standards, though conditions are often enhanced by the allocation of the unlimited free use of a company car and other benefits associated with the job. Engineers particularly suffer from low pay and lack of status. Lawyers, medical practitioners and accountants are often regarded as higher in the social scale and salaries are a fraction those in France and Germany.

Despite a temporarily falling market, the British construction industry cannot be ignored by company strategists. British companies are encountered bidding for work in other Member States. The British style is to 'muddle through' and 'get there in the end'. The British construction industry has certain strengths and has proved itself competitive in world markets.

SPAIN

Spain, the second largest country of the Community and twice the size of the UK, occupies most of the Iberian Peninsula leaving a sixth for Portugal, its western neighbour. Its population of approaching 40 million compares with Portugal's 10 million. Spain has the highest unemployment rate in Europe, 20.1 per cent in 1987, while Portugal's was only 7.0 per cent. Despite being close neighbours, Spain and Portugal are separated by historical, cultural and language differences, as well as by a chain of mountains.

In the last 15 years Spain has adapted from General Franco's dictatorship with strong central structure, isolated from the rest of Europe, to a pluralist democracy allowing considerable autonomy to the regions. On 1 January 1986, Spain's destiny shifted considerably when it joined the EC.

Spain shares with Italy a poor record of implementing EC directives. This is partly explained by the country having to digest a large number of existing directives when it joined the Community, though some would attribute it to the well known *vuelva-usted-mañana* (come back tomorrow) attitude of the Spanish administration. The implementation process is also slowed by the quasi-federal structure of the country (though not strictly a federal state, Spanish regions have similarities with the German *Länder*). The regions, particularly, find some directives difficult to implement.

Spanish social values are changing. Machismo is losing ground, more women are joining the work force and a few are bold enough to throw themselves into careers. Making money is in vogue and the 'yuppies' abound in the main cities such as Madrid (which lies exactly half-way between Rabat and Paris) and Barcelona. However, some aspects of Spanish lifestyle have not changed much. The traditional long lunch and the siesta show no signs of disappearing. However, in compensation the business community are prone to work late into the evening.

The Spanish are generally well educated, appreciate culture and are hard working. Above all they are a proud people, but nevertheless open and welcoming to foreigners. Since 1975 many young people have gone abroad for higher education, primarily to France and the UK. This new generation is sharply divided from the generations moulded in Franco's era, particularly in terms of ideology and business. The new wave is sharp and well aware of the business opportunities.

The regions

The regions differ profoundly. As well as differences in climate and terrain, the 17 regions differ in culture, industry, size and administrative structure. Some of the regional languages, such as Catalan and Basque, are very important and bear little relation to Spanish. The 1978 constitution transferred many powers from the central authority to the regions. Through regional government and parliament, each region enjoys certain legislative authority. Over the last few years this has been exercised by some of the regions to introduce many changes in the field of construction and planning, amongst others. Some regions such as Cataluña and País Vasco (the Basque Country) have brought in numerous regulations, while others have left the general framework largely untouched.

The 8000 or so local authorities, grouped into 52 provinces, unfortunately do not all appear to have the resources needed to fulfil adequately all their functions, particularly in relation to planning and construction.

Construction

After the car industry, construction is the sector most targeted by foreign bidders. This is not surprising since Spain is the star performer in European construction in terms of growth. Its market is the fifth largest in the Community and is enjoying a prolonged boom which is likely to continue despite official concern about inflation, the highest in the Community. Table 13.6 comparing the characteristics of the workload with that of the UK shows marked differences, with the high levels of public work and the low level of repair work indicative of

Table 13.6 Comparison of construction output in Spain and the UK 1987

	Spain £ billion	UK £ billion
Housing	5.91	6.75
Commercial	7.23	8.44
Public	6.50	3.88
Repair	3.17	15.54

Source: Euroconstruct

a country in the growth stage of economic development, whereas the UK could be described as in the mature or even crumbling stage.

Growth

Spain has the fastest growing economy in the continent of Europe which makes it attractive for foreign investors despite the problems, mainly of a regulatory and fiscal nature, of getting established there. Economic deregulation accompanied by a flood of foreign investment, 85 per cent of which originates from other Community countries, and massive EC funding programmes have kept the once under-utilised construction industry very busy.

The construction industry embarked on a growth surge in the mid-1980s (see Table 13.7 for sector growth rates 1985-89). This peaked at a spectacular 13 per cent real growth rate. The main increase was in public investment expenditure which far exceeded private sector spending. The growth rate has since slowed to around a still impressive 10 per cent. Not unnaturally, the demand for construction has brought a shortage of skilled labour and rising costs. Labour costs grew by 10 per cent between December 1988 and November 1989.

A number of special projects are giving a sparkle to Spanish construction. The year 1992 marks the fifth centenary of Christopher Columbus's voyage to the New World and is being celebrated by a specially housed exhibition. The Olympic Games take place in Barcelona in 1992 and the facilities for Expo 92 are being built on an island on the Guadalquivir River in Seville, Andalucia, Spain's largest and poorest region. The latter is the country's largest ever single civil engineering project and includes 90 country pavilions. The scheme has created 2000 jobs during construction and the Expo State Corporation

Table 13.7 Growth in construction in Spain

	1985 %	1986 %	1987 %	1988 %	1989 %
House-building	1.0	5.0	7.0	5.4	5.0
Non-Residential Building	1.0	6.0	9.0	9.3	15.6
Civil Engineering	2.0	3.0	4.0	20.0	25.0

Source: SEOPAN Report to European Federation of Construction Economists

is planning to convert the site after the exhibition to a high technology city, which will create an estimated 8000 permanent jobs, many of which would be for scientists and engineers. Companies which are planning to take up space on the site include IBM, Fujitsu, Philips, Coca-Cola and Rank-Xerox. Companies are being offered an 85 per cent discount on taxes against construction costs and equipment if they stay on after 1992. The year also marks the transitional stage to full membership of the Community.

1989 was a particularly prosperous year for construction, with a 13 per cent increase on the previous year. Even so housing construction could not outpace growing demand, up to 30 per cent of which was from external sources such as UK nationals seeking second or retirement homes. Demand is likely to continue, but at a slightly lower rate. Non-residential building grew at 15.6 per cent in 1989, with the public sector proving particularly vigorous, with civil engineering increasing by 25 per cent.

With residential building growing less sharply, the civil engineering sector is expected to continue to grow rapidly. Growth approached 20 per cent in 1990. Demand will continue strongly, and it will take decades to bring the infrastructure up to the standard necessary to meet the requirements of the population and economy as it responds to the opportunities of the European single market. Apart from the showcase work for 1992 a massive civil engineering effort is required to fulfil the *plan de carretas* (General Road Plan), the *plan de transporte ferroviario* (Railway Transport Plan), port and airport programmes and land reclamation and water distribution schemes.

Structure of the construction industry

The industry structure follows the familiar pattern of a large number

Table 13.8 Output of the major Spanish construction companies in 1987

Company	Domestic	Foreign ECU million	Total
Dragados y Construcciónes SA	850	161	1011
Agroman Expresa Constructora SA	540	83	623
Huarte y Cla	424	52	476

of small local firms and a small number of large firms. The large construction companies (see Table 13.8) must not be underestimated. They are strong and have a history of successful export operations, particularly in Latin America and the Middle East. Large-scale contractors operate nationally and tend to have a civil engineering background. Few of them get involved in building development.

Ownership

Ownership patterns of the major construction groups are extremely complex. Ultimate control is generally held by large banks and financial holding companies. This is mainly due to the Spanish system of corporate finance which is based on powerful banks and a relatively weak stock market. Many of the groups are intimately involved with non-construction activities such as manufacturing, testing and research, motorway operation, car parks, and food and service industries. As these firms restructure to face the competitive pressures of the single market, changes in ownership are likely.

The banks control the contractors and also get closely involved with project finance for the developers. Accordingly they are able to influence project selection and development as well as contractor selection. Foreign contractors are strongly advised to arrange their own financial backing.

Investments

Foreign investment plays a major role in the economy in providing not only finance and technology but also management know-how in a country sorely lacking professional skills.

Management

Management principles applied in Spanish firms differ largely from those found in similar sized organisations in northern Europe. For instance, in Spanish companies team work, if part of the vocabulary at all, is understood to mean individuals working independently under a strong leader. Leadership in Spain has overtones of machismo and bravado. The strong but benevolent autocrat is regarded as the ideal leader. Openness and discussion of problems is not part of the management culture and communication between departments may be minimal.

Values also differ from the European norm. For example, there is pride in coping with the permanent crisis and emergency which reign in Spanish businesses. Forecasting and planning are not part of Spanish management style. Who knows what tomorrow will bring? Reporting, monitoring and interpretation are generally carried out but lack the consistency needed for proper analysis and planning. Intellectual schemes and grand designs so loved by the French are not even dreamt of.

Family-owned businesses still predominate, and are usually run by a single leader with almost no middle management, even in companies employing as many as 500 people. They also tend to be highly compartmentalised, bureaucratic and run on authoritarian lines.

Particular caution must be shown in using company accounts to assess the health of Spanish small and medium companies. Turnover and cashflow figures are particularly likely to be suspect. There are generally several sets of figures, one for the bank, one for the taxman and (why not?) one for the family. The reason for this is that the accounting profession in Spain is at an early stage of development. There is no recognised standard practice.

Those wanting to set up in Spain will encounter interminable administrative delays with convoluted procedures whose original purpose has been long forgotten. Disillusioned old Spanish industrialists are likely to advise those with business ambitions to spend a couple of years in the civil service simply to discover the loopholes and shortcuts and to make strategic acquaintances.

The reality is that the *intimo amigo*, the very close friend, is a vital attribute to succeed in Spain. Sadly, one's relations are often more important than one's skills. For promotion one needs personal loyalty, friendship and ability, in that order.

Main organisations

Two ministries are principally involved with construction. The first is the Ministry of Public Works and Planning (MOPU) which deals with construction regulations and standards, roads, subsidised housing and ports. The second is the Ministry of Transport, which is primarily concerned with airports and railways.

Architects and engineers

Architects and engineers enjoy high status. Their professions are highly protected and regulated and are legally bound to their *colegio*, their 'professional colleges'. Their respective services are obligatory for building and civil engineering construction. In consequence there is no design and build activity, but contractors generally take responsibility for much detail design. Architects are wholly responsible for ground investigations (geotechnical investigations are compulsory for all construction projects), the design of structures, electrical, heating and air conditioning installations and site supervision. Their tasks and fees are fixed by royal orders.

The peculiarity about the Spanish system is the part played by *colegios*. All construction projects must be examined by the regional *colegio*. There are 17 *colegios oficiales de architectos* and two main *colegios de ingenieros superiores*. The *colegio* checks the quality of the project on behalf of the authorities before application is made for planning approval and the tendering procedure started. The *colegio* also collects fees, retains a fraction and then pays the remainder to the architect. *Colegios* only recognise individuals, not firms.

The provincial *colegios oficiales de aparejador*, colleges of 'assistant' or 'technical' architects, have a parallel and similar part to play to that of the regional colleges of architects. *Aparejadors* have no executive responsibility but provide technical direction on site. By law they have to be appointed on each construction site. In some respects their functions can be compared to those of quantity surveyors in the UK.

Civil engineers and contractors

Civil engineers design and supervise work in the same way as architects, frequently working in partnerships of five to ten. Many are employed by design offices affiliated to construction companies.

Clearly a design office owned by such a company has little power when it comes to site supervision. Given the relationship between banks and construction companies it can also result in a situation where the contractor, and associated designer, are selected according to the contractor's financial connections rather than on merit or price.

In Spain a construction contract is almost invariably entrusted to a general contractor who assumes all responsibility. There is generally no transfer of liability to sub-contractors. Contracting remains a low-margin business (1 or 2 per cent), despite the high level of demand.

On 95 per cent of contracts the contractor is approached only once the design is complete. Estimated quantities (similar to bills of quantities in the UK) are prepared by the technical architect and issued to tenderers to be priced and returned. General contractors usually get paid for work as measured, and rarely on a lump sum basis.

Standards

In contrast to Italy, building regulations generally stem from national level, though some regions have made use of their powers to make their own regulations.

The legislation relating to building regulation is divided into two classes. The *normas basicas de la edificación*, basic norms or NBEs, cover aspects such as structural steelwork, fire protection, seismic design. These are rules produced by the former Housing Ministry and are the only obligatory standards. The second category is the *normas tecnológicas de la edificación*, technical norms or NTEs which remain optional, even for public works. They are prepared by the present Ministerio de Obras y del Urbanismo or MOPU.

MOPU publish a useful guide to the building regulations known as *Índice de Disposiciónes Relaciónadas con la edificación* (guide to provisions relating to building). Another valuable professional guide to construction is *Guía Profesional de la Construcción* (1985) published by the Spanish building employers' confederation, Confederación Nacional de la Construcción, which describes the main legislation.

For product standards, the Spanish equivalent of the British Standard Institute is the Asociación Español de Normalización y Certificación, AENOR. Spanish standards, *Una Norma Española* or UNE, deal with materials, product specifications and testing methods. Little attention is generally paid to conformity to the national standards. Products are imported without regard to compliance to UNE. However, the installed product has in the past not been subject to insurance nor has the supplier carried any liability. Clearly the implementation of the product liability directive (85/374/EEC) will change matters, as would other Commission proposals on liability.

Contracts

There is no restriction on the terms of private contracts provided the civil code is complied with. Neither do restrictions apply to contracts for subsidised work, though the work must comply with the state's requirements. Government contracts and the procedures for their award are subject to legal requirements and a regulation issued by the Ministry of Housing. The Public Works Directive (71/305/EEC) does not come into force until 1 March 1992, and even then some minor provisions are further deferred.

Liability

The Spanish Civil Code, which follows the patterns of the *Code Napoléon*, prescribes a 10 year strict liability for serious defects. This applies simultaneously to the architect, the technical architect and the main contractor. The architect in practice bears most of the responsibility and unlike the contractor carries an insurance policy. Should a defect occur within 10 years of completion of the works, the owner may embark on proceedings to obtain the cost of repairs until 15 years after discovery. For engineers who design structures, anything which does not comply with the contractual specification is treated as a defect. A compulsory five-year guarantee applies to private developers who construct subsidised housing for low-income households. Suppliers of materials and components, including sub-contractors, have no liability to the owner. Nevertheless there is a trend for liability to be passed on jointly to sub-contractors.

Conclusion

Spain as a place to work has great attractions. Expatriate wages generally match international norms. The normal working week is limited by law to 40 hours. There are 14 public holidays and 30 days entitlement to statutory paid annual leave. Public holidays are not moved to a Monday as is done for most 'bank holidays' in the UK. The practice of taking leave days between a weekend and a midweek public holiday is known as bridging. In one month there are so many public holidays that the bridges are sometimes linked up into a 'viaduct'. Northern Europeans take readily to the long lunches and siestas in the hot dry summers (so hot that Spaniards actually prefer to spend their holidays on the cooler Atlantic side rather than on the Mediterranean coasts). Apart from the traffic congestion in the cities and an appalling telephone system, Spain can be close to idyllic.

THE NETHERLANDS

The Netherlands has a population of 14.7 million and the highest population density in the Community with 356 per square kilometre. Half the population lives in a narrow low-lying urban arc called the *Ranstad* or 'city on the edge'. The capital, Amsterdam, lies at one end of this strip and Rotterdam at the other. Europort, the world's largest and busiest port, is the port of Rotterdam. Within this strip is The Hague (den Haag), the political and diplomatic centre.

The Netherlands is renowned for its dykes. It is the flattest country in Europe and much of its area lies below sea level and has been reclaimed. The Dutch have also become famous for their relaxed attitude to drugs and prostitution. Drugs are not legal, but the policy of tolerance, which extends to the free provision of needles to addicts, appears to have been more successful in controlling the problem than more repressive approaches seen in other countries.

Other aspects of social policy are also remarkable. There is political consensus for attempting to maintain a link between private sector wages, and wages for public sector employees, pensions and social welfare benefits. Recent governments have had to ease this link as pressure has grown on finances. The disability scheme (WAO after its name in Dutch) accounts for almost 7 per cent of Gross National Product and the numbers supported by it, 100 000 of whom are under 35 years of age, are approaching one million. A large percentage of those living on WAO are eligible on psychological grounds, often due to stress at work or nervous exhaustion. The unemployment benefit regime is also liberal in operation and in the benefits offered.

The environment is a major concern amongst politicians and the public. Recent episodes, such as periods of smog, dying forests, beaches polluted by algal bloom and many seals dying in the North Sea have added to concern about nitrate and pesticide residues in drinking water and contaminated soil. Contributing factors to the problem are population density, a density of cars per square kilometre almost double that of the UK and 50 per cent more than that of Germany, and an intensive agricultural sector which results in nitrogenous fertiliser application nearly two and half times that of the UK. Much of the airborne pollution is generated in surrounding countries and pollution in the major rivers, such as the Rhine, is mainly upstream. The government response is the National Environmental Policy Plan (NEPP) tabled in May 1989, subsequently modified and known as NEPP-plus. This plan, probably the most ambitious in any country, aims to integrate '. . . environment and economic decision making, as a means of contributing to sustainable development'. The policy covers waste disposal and recycling, agriculture, energy use, transport and other forms of pollution. Since 1988 a general environmental levy on fuels has been payable, generally in line with the relative carbon content of the fuels. Transport policy proposals to deal with pollution and congestion have recently been considered, but attempts to discourage car use by making it more costly have provoked outrage amongst some drivers. The proposals have attempted to spare the transport industry which is one of the largest in the Community.

BElgium, the NEtherlands and LUXembourg have for 30 years co-operated in the BENELUX Economic Union. The BENELUX countries together with France and Germany have created under the Schengen accord a borderless market in which goods and people can move freely. The Netherlands has benefited from membership of the Community and strongly supports further integration. The giving up of national sovereignty is not controversial. The Dutch attitude to Europe is a product of its export-led economy. The educational system puts a strong emphasis on languages, with ability in English and German usually excellent amongst professionals. French and Spanish may also be known.

The economy

The Netherlands is highly dependent on foreign trade and the economy is correspondingly open. The value of exports of goods and services was 55 per cent of Gross National Product in 1988, while imports were equivalent to 50 per cent. The major trading parter is Germany, followed by Belgium and Luxembourg (together as the other BENELUX partners), the UK and France for exports and the USA and UK for imports. In many respects the Netherlands is a good gateway into continental Europe. Growth has remained steady at around 3 per cent and is likely to dip a little in 1991. Inflation has remained very low, 0.6 per cent in 1988, 1.1 per cent in 1989 and was predicted in 1990 to remain below 3 per cent in 1990 and 1991. Unemployment is around 6 per cent of the potential workforce.

Construction activity

Approximately 5 per cent of EC construction takes place in the Netherlands. In 1988 repair and maintenance was the largest sector, accounting for over 40 per cent of output. In 1989 construction output was 5.9 per cent of GDP, and employment in construction was 7.4 per cent of civilian employment.

The short-term prospects of the Dutch construction industry are moderate. The year 1990 saw no growth, with residential investment falling sharply. The Dutch construction industry is facing rising construction costs and shortages of skilled labour.

Statutory public holidays are six days per annum with an extra day every five years. Statutory paid annual leave entitlement is four weeks, but collective agreements usually increase this five or six weeks. Salaries are around 40 to 50 per cent higher than those in the United Kingdom.

BELGIUM

Belgium is a country of 30 500 km² and 10 million inhabitants contained between other Member States and the North Sea. Belgium is noted for being enthusiastically European, which is fortunate as it houses a major part of all the key Community institutions as well as the North Atlantic Treaty Organisation (NATO or OTAN in French) and other international organisations. It is one of the founder members of the Community.

Belgium is a relatively small country split into two main communities – the Walloons, who speak French and are traditionally the upper class, and the Flemings, who speak Dutch and are currently doing better economically than the once more well-to-do Walloons. French, spoken predominantly in the southern part of the country, Dutch (sometimes referred to as Flemish, even though the Flemings refer to it as Nerlandais; it is as close to Dutch spoken in Holland as English is to American English), spoken in the north and west, and German, which is spoken in the eastern corner of the country, are all official languages.

Bruxelles, Brussel or Brussels, depending on where you come from, is the capital and also the largest city with nearly one million inhabitants. The capital is the only part of the country which is officially bilingual (French and Dutch), although with the presence of international organisations such as NATO, English is also widely spoken. The public transport system is superb with a spotless *métro*.

This is integrated with the buses and trams, and a BF35 ticket (55p) gives complete freedom to use any of these modes of transport in any combination for up to an hour.

However, for motorists Belgium can be harrowing. Belgium retains the *priorité à droit* rule in spite of all evidence of the unnecessary danger involved. (This rule requires a driver to give priority to all traffic coming from the right.) Insurers automatically, and irrevocably, assume that damage to the left-hand side of another vehicle involved in an accident is your fault. This, added to the fact that a driving test has only relatively recently been introduced for new drivers, means that accidents are frequent. Accident statistics are apparently kept down (although they are still high) by only recording a road death if death is at the scene of the accident. In compensation, the police appear to take no notice of where cars are parked unless the commune happens to be running short of funds.

The climate is similar to that of the UK though damper and with colder winters. Belgium is a country of idiosyncrasies, with superb organisation in some respects and unbelievable lapses in others. The Atomium, the giant representation of a molecule and symbol of the twentieth century, had as late as 1989 (and probably still has) displays including yellowed, curly photographs of 'modern technology' which date back to when it was opened in the early 1960s, and perspex windows rendered opaque by time and scratches.

Second-hand cars are subject to VAT at the rate of 25 per cent of the imputed value, which means that the value of cars plummets. Houses also attract a tax, though at a less penal rate, on transfer of ownership. Even in the centre of town there are still empty plots, though the property market has been moving in recent years with many companies wanting to locate offices in the 'capital of Europe'. Much of the commercial property has been bought up by Swedish interests.

The economy

The Belgian economy is small and open, making it sensitive to the external economic climate. Exports and imports in 1988 were each worth nearly 60 per cent of GDP; the UK figure is about 17 per cent. GDP per head at $15 180 in 1988 was above that for the UK ($14 413), although when allowance was made for purchasing power parities the UK came out ahead. Even so, there were rather more cars per capita in 1988 in Belgium than in the UK. Growth in GDP was relatively sluggish in the period 1983–87 but climbed to 4.2 per cent in 1988 and has stayed above the Community average since then. Growth was 4 per cent in 1989 and 3.6 per cent in 1990. OECD forecasts in February 1991 were for 2.3 per cent growth in 1991 and 2.6 per cent in 1992, in both cases slightly above the Community mean.

Unemployment is relatively high at around 9 per cent; inflation is low and stays close to German levels. Rates of interest are also generally closer to German than to UK levels. Unemployment is one problem; government debt is the other main problem – at 120 per cent of GDP it is the highest in the Community except for Ireland. The government spends heavily on labour market programmes; in 1988 it was equivalent to 4.2 per cent of GDP. One programme is a system of income enhancement for those taking early retirement. In 1988, 133 000 or 3.5 per cent of those employed benefited from this scheme. Half of all those in part-time employment were receiving unemployment benefit. As in Denmark, there is no limit on the duration of unemployment benefit. However, government spending at 49 per cent of GDP is not quite at the levels of Holland and Denmark.

Construction

An unusually low proportion – 5.6 per cent – of those employed in Belgium work in the construction industry. The market is much smaller in absolute and relative terms than in the four largest Community countries. It is nevertheless still substantial. Private residential and private non-residential construction grew steadily in the six years to 1990. Rehabilitation and maintenance work also grew. Unfortunately, the high levels of public debt referred to above cause the public sector to be periodically shaken by budgetary crises which usually lead to severe restrictions in public investment. The 1990 situation may be described as disastrous with Belgium ranked bottom amongst the 12 Member States in terms of public investment as a proportion of GDP. Cuts in public investment expenditure have left large civil engineering firms with underutilised capacity and their workload has halved since 1980.

Structure of the construction industry

A few large companies (see Table 13.9) make up the backbone of the construction industry. Only 13 companies employ more than 500 people compared with 40 in the Netherlands, a similar sized country. The rest of the industry comprises a multitude of small local firms. Contracting abroad is rare with firms being inwardly orientated.

Table 13.9 Output of the major Belgium construction companies in 1987

Company	Domestic	Foreign ECU million	Total
Les Entreprises SBBM et Six Construct	190	121	311
Companie d'Entreprises CFE	137	81	218
S.A. Franki NV	41	115	156

Context

Belgium has a written law of 'Latin' inspiration and a fragmented political structure. The country is divided into nine provinces which are also district authorities. These exercise some state delegated power over the 2500 or so local authorities – the *communes*. Each province and *commune* has legislative powers.

The main state organisation dealing with construction is the Ministry of Public Works which retains the key decision-making powers, particulary in relation to land use, planning, major transport infrastructure and state building.

Generally speaking, the state does not propose much innovation in the fields of construction law or technology except perhaps in cases when it acts as a client. The state is nevertheless seen as a leading player and is often referred to in jurisprudence as a model of good practice.

Standards

There are numerous standards, *normes*, but only a few have been certified, *homologué*. Standards are usually given effect under the construction contract. Approvals, *agréments*, on the other hand are full parts of the standardised technical specifications, *spécifications techniques unifiés*. In the absence of standards or approvals other types of proof of quality of a product will be considered.

Controls

The state bears responsibility regarding regulation on building work, safety at work, fire prevention and public safety, but direct intervention is rare. As in Germany there are no government building inspectors. Supervision is delegated to local and regional levels. There are no national regulations except in fields such as fire safety of tall buildings and public buildings.

In practice, responsibilities are handed over to the client and the builder is obliged to observe construction laws and regulations. Contractors are required to provide proof of compliance with regulations on demand from a government official. In this sense there is permanent self inspection within the building team.

Nonetheless, there are private technical inspection offices, *bureaux de contrôles techniques* as in France which are meant to prevent risk of damages. Supervision by one of these organisations is required if 10 year liability insurance cover is to be provided.

Architects and other construction professionals

In Belgium, there are basically three parties, the client, the architect and the contractor. Oddly enough, the law has never acknowledged the existence of other participants such as design offices, quantity surveyors, planners, decorators or landscapers, although their services are widely used in the industry. Even engineers are almost non-existent in the Belgian legislative context.

Architects in Belgium have a professional monopoly. Only they have the authority to draw up and sign the documents attached to applications for planning permission. This protected position is further reinforced by an obligation to decline any work for which they are not retained for supervision, unless assured that another architect will have responsibility for that task.

The responsibilities of an architect are considerable and include the stability of the building, even though an engineer may have been called in. An architect may be made responsible by a client for carrying out any of the activities involved in the construction process, but exclude acting as a contractor.

Architects by law have a major role to play. They are obliged to defend the public interest as well as the client's. They may also be required to act as arbitrator. However, it appears that these concepts are becoming out of date. Written agreements defining respective roles and responsibilities for other participants are diminishing the architect's pre-eminent position. The trend is for architects to restrict supervision of work to 'general direction' and avoid 'permanent supervision'.

Contracts

No standard contract is imposed by the order of architects but various forms of contract are available. They naturally define the task, the obligations of the architect and client.

Regulations for state contracts are not as codified as in France but are highly structured. Government construction contracts are always given out to a principal contractor whose sub-contractors may not be known to the client. From a legal point of view, the principle contractor is seen as a client to his sub-contractors, who are thus also subject to a 10-year liability. No text defines a framework for sub-contracting.

Liability

Generally speaking, the contractual relationship provides for a 10 year period of liability. This applies to contracts between contractors and sub-contractors but not between contractors and suppliers. An engineer or design office who is bound by contract to the architect rather than the client is not subject to the 10 year liability period.

Ten year liability also applies to civil engineering structures such as bridges, drains, industrial structures, water towers and quay walls.

Independently of the 10 year warranty to the client, runs a semi-criminal liability, similar to tortious liability in the UK. This continues for up to 30 years from the act or default which caused the loss. In this context, clients can sue sub-contractors or suppliers directly.

Insurance

Belgian law does not require insurance of the object built. Only architects are required to have professional indemnity insurance.

Supervisory insurance is always taken out for a given site. This covers all builders working on the site and continues in effect for 10 years. It also acts as the builder's liability insurance and costs about 2 per cent of all the work done. It should be noted that 10 year insurance can only be obtained if design and construction are subject to technical inspection by a specialist.

Conclusion

In Belgium, a working week is 40 hours, the lunch break is spread over at least two hours and statutory public holidays and paid annual leave amount respectively to 10 days and 24 days.

DENMARK

Much of Denmark is made up of islands. About 30 per cent of the population of 5.1 million live in the capital, Copenhagen, which is on Sjælland (sometimes referred to in the UK as Zealand), the largest island. In the Community, only Ireland and Luxembourg have smaller populations. Its low population density of 119 per square kilometre is similar to that of France. The Faroe Islands and Greenland are part of the Kingdom of Denmark. The Faroe Islands are a collection of 22 islands, 18 of them inhabited, in the North Atlantic between Scotland, Norway and Iceland. The population of 44 000 is largely dependent on fishing. Greenland is independent in most respects and withdrew from the Community in January 1985. The population is about 50 000.

The Danish sectors of the North Sea produced about half of the country's oil and gas requirements in 1986. Another source of energy, wind, is the subject of much interest in Denmark, which intends to produce 10 per cent of its energy from this source by the year 2000. There are about 3000 wind turbines, mostly owned by private co-operatives. Only California has more. One recent development involves the construction of 11 wind turbines on reinforced concrete bases between 1.5 and 3 kilometres offshore in up to 5 metres of water.

Taxes are high, both on income and VAT. The rates of income tax, between 40 per cent and 22 per cent, are not in themselves high, but the relative lack of allowances and exemptions result in the effective tax (tax revenue as a percentage of the tax base) being the highest in Europe at 26.3 per cent. The effective personal income tax for the UK, Spain and Germany are 11.5, 10.8 and 6.5 per cent respectively. The standard rate of VAT of 22 per cent is high, as is the effective rate at 18 per cent. The standard and effective rates are 15 and 9.4 per cent in the UK (standard rate increased to 17.5 per cent in the March 1991 budget), 12 and 6.4 per cent in Spain and 14 and 10.7 per cent in Germany. Company tax was reduced to 40 per cent in 1990. Much of the Danish tax revenue is used to support a generous social security system, which enables some to 'retire', a few even before starting work. Unemployment payments are high at 90 per cent of earnings, and do not reduce over time. Employees make no social security contributions, but employers' contributions, amounting to about 3 per cent of salaries, are payable. Government expenditure is accordingly high at nearly 54 per cent of GDP. This figure is almost the same as for the Netherlands, another country with a very generous social security system, but is much higher than other Member States. The corresponding figures for the UK, Germany and Spain are 43 per cent, 43 per cent and 36 per cent respectively.

The Danish language is one of the Nordic group of languages and has strong similarities to Swedish and Norwegian. In general Denmark is pulled in two directions – towards the Scandinavian north, and south towards the Community.

Denmark and the Community

Denmark joined the European Communities in 1973 at the same time as the UK. The agricultural sector, which is an important contributor to the economy, has benefited from Community membership and has generally been rather more in favour than the rest of the population. The Danes seem only recently to have stopped regarding themselves as alien Scandinavians in the Community. The breaching of the Berlin wall may have contributed to the change of attitude. Since then the government has spoken strongly in favour of European integration, possibly seeing it as a counter to the threat of German dominance.

The economy

The standard of living in Denmark is one of the highest in Europe with gross domestic product per head in 1988 of US $20 972. In 1987 the GDP per head was higher than that of west Germany, even when the higher cost of living was taken into account. The economy experienced a recession in 1986 and growth has remained sluggish since then and is likely to remain modest in the short-term. Prices have remained fairly stable with consumer price inflation around the 5 per cent mark. Unemployment has stayed at around the 9 per cent level for several years.

Construction activity

Danish construction represented about 3 per cent of the EC construction market in 1989. In 1988 183 000, or about 7 per cent of total employment, worked in construction. This figure represented a 2 per cent fall on 1987 and the decline was accelerating into 1989. The output of the industry declined by 2.5 per cent in 1988, 4.3 per cent in 1989 and 1.1 per cent in 1990 and the global recession appeared likely to prolong the period of belt-tightening for Danish construction companies. There remains over-supply of office space in Copenhagen, with little new construction, particularly outside the central area. Companies specialising in commercial and industrial building have been hardest hit.

The only growth sector is public works, and this is largely due to the massive Storebælt project. This 18-kilometre-long crossing is worth £1700 million and is expected to be complete by 1996. The 6.5-kilometre, £300 million western bridge, a multi-span combined road and rail low-level link, was started in 1988. The original conditions for this contract contained the requirement to use, to the maximum extent possible, Danish labour, materials and sub-contractors. These conditions and the conduct of the tender procedure were the subject of a complaint to the Commission which was discussed in chapter 6. The 8-kilometre eastern link had to allow 65 metres clearance for the busy shipping lanes. While a bridge solution involving a 1624-metre suspension bridge was chosen for the road link, a twin bored tunnel was chosen after a design competition for the rail link. The Danish consultants COWIconsult have been the lead consultants for the project with Mott MacDonald involved for the design of the tunnel. Contractors include Monberg and Thorsen of Denmark, Ballast Nedam of the Netherlands, Jan de Nul of Belgium, and Taylor Woodrow of the UK. In February 1991 the project was 10 months behind schedule, in part due to delayed delivery of the tunnel boring machines being manufactured by the Scottish firm James Howden. The rail link is planned to be complete three years earlier than the motorway and will provide a fixed link between Copenhagen and mainland Europe.

In March 1991 the Danish and Swedish governments signed an agreement for another major construction project, a 17-kilometre link across the Oresund, the channel that divides Sjælland (and Copenhagen) from Sweden. This motorway-rail bridge is estimated to cost DKr 12 billion (about £1.06 billion or ECU 1.5 billion) and is due for completion around the year 2000. This bridge, together with the Storebælt and other links, brings the investment in transport infrastructure in the region to about £5.6 billion (ECU 8 billion) in the 1990s.

Construction professionals

There are no legal restrictions on the practice of engineering or architecture in Denmark, but those practising are normally members of the Danish Association of Consulting Engineers (Foreningen af Rådgivende Ingeniører) or Danish Association of Practising Architects (Praktiserende Arkitekters Råd), both of which have strict admission requirements. There is no official regulation on tendering procedures, but for design work competition is rarely on price. On large contracts consultants may be required to submit technical proposals, while competitions are often used for the selection of architects.

PORTUGAL

Portugal has a population of 10.3 million, which is 3 per cent of the EC population and makes it a similar size to one of the provinces of Spain. It is relatively sparsely populated with 112 people per square kilometre. The largest cities are Lisbon with a population of about one million and Oporto with about 700 000. Oporto, where umbrella sales are strong due to the heavy rainfall in winter and tendency to damp and foggy weather, is enjoying a new prosperity and is benefiting from an improved rail link with Lisbon. The country has had a 600-year association with Britain, with whom the country shared a common hostility towards the Spanish. One of the effects of Community membership is to break down the barriers between Portugal and its large eastern neighbour with the impartial imposition of the various freedoms of the Community and harmonisation measures. The Spanish are now the largest foreign investors in Portugal and accounted for over 12 million visitors to the country in 1989. Only a tenth that number came from the UK, the next most important source of visitors. Spanish and Portuguese are similar languages in the written form but sound very different. Spanish is easier for a Portuguese speaker to understand than vice versa, since Portuguese pronounciation is more compressed. The cost of living for expatriates in Lisbon is similar to that in London, but away from the city it is much less expensive. The roads are the most dangerous in the Community.

The islands of Madeira and the Azores are administratively part of Portugal. The Madeiras, about 800 kilometres southwest of Lisbon off the coast of north Africa, have a population of around 280 000 and are a popular year-round tourist resort. The Azores are a scattered group of islands about 1200 kilometres to the west of Lisbon, inhabited by about 290 000 people. The climate is wet in winter with high winds. Summers are hot and humid. There is interest in developing the geothermal energy resources of the Azores.

Economy

Portugal and Greece are the two poorest countries of the Community in terms of GDP per capita. The unemployment rate, which is running at about 5 per cent, is rather better than Spain's (around 20 per cent). Inflation is around 14 per cent and likely to stay at that level for some time. Unlike Spain, Portugal has yet to join the European Monetary System. The relatively large but very inefficient agricultural sector employs about 18 per cent of the workforce but contributes only 7 per cent to GDP. Shoes and textiles account for about 30 per cent of export earnings but the industries are likely to suffer from more open competition and become subject to restructuring. Shipbuilding and ship repair facilities are noteworthy, with the Lisnave repair facilities being the largest in the world.

When the authoritarian regime of Salazar ended with the 1974 revolution, the banks and many of the country's major industries were nationalised. In April 1990 privatisation legislation was adopted and in 1991 the proceeds of privatisation are expected to equal 2.5 per cent of GDP and to improve general economic performance. GDP growth remained above 4 per cent between 1984 and 1988. In 1989 it reached 5.4 per cent but in 1990 dropped to 3.9 per cent (preliminary estimate). OECD figures prepared in late 1990 predicted real GDP growth to continue at around the 4–4.5 per cent range until at least mid-1994, with unemployment heading downwards to 5 per cent.

Portugal and the Community

Portugal joined the Community on 1 January 1986. The population and the business community are strongly committed to membership, though the original enthusiasm has given way to a more wary attitude. Portugal has benefited considerably in financial terms from membership of the Community. For the purposes of the European structural funds it is an 'objective 1 area', i.e. a region whose development is lagging behind and where the per capita GDP is less than or close to 75 per cent of the Community average. Structural funds to Portugal amounted to 2.4 per cent of GDP in 1988 and 1989 and rose to 2.7 per cent in 1990. In 1989, ECU 850 billion were received, or about 11 per cent of total domestic investment. The funds have gone into roads, railways and other infrastructure as well as into agriculture and professional training. The improved communications have reduced regional imbalances and industrial development has been stimulated. A number of major foreign manufacturers have chosen to take advantage of the low wages and recently liberalised labour market by setting up facilities in Portugal. These include Ford, General Motors of the USA and the Valmet Corporation of Finland. The government provides incentives for foreign investors including those from outside the Community. The effects are visible: new factories are springing up in Portugal, older buildings are being refitted.

Portugal is on the western periphery of the Community and has become concerned that developments in eastern Europe will reduce its attractions for investors as well as diverting Community funds in that direction.

Construction

The Portuguese market presents excellent opportunities in both civil engineering and building despite its relatively small size. Construction represented 5.7 per cent of GDP in 1987, but has grown to 7.9, 10.9 and 8.0 per cent in the years 1987, 1989 and 1990 respectively. In comparison with the larger countries output is tiny, but the continuing boom in investment in infrastructure and other development gives the market attractions.

The majority of foreign investment has come from within the European Community. Investment rose 22 per cent in 1987 and that chiefly for infrastructure projects. In 1989 a detailed regional development plan was submitted to the European Commission in connection with an application for structural funds. This plan set out ambitious targets for infrastructure, communications and industrial development. Years of neglect left Portugal with an inadequate and patchy transport system. Some roads, even in the main cities, are in poor repair. The country is also short of airports.

Large developments for tourism have been undertaken throughout the country to cater for growth in this sector. Planning regulations related to tourist development have been lax in the past but have recently been tightened. The procedures can be slow and bureaucratic, particularly with increased pressure for development combined with concern about the environment and the insensitive nature of many developments constructed in the 1980s.

As with other countries whose legal tradition stems from the *code Napoléon* the roles of the professionals are defined by law, though in Portugal the professions are less regulated than in Spain and foreign practitioners should be able to adapt without great difficulty. There is a legal requirement for both architects and engineers to be registered if they are to practise in Portugal. Many aspects of building design could previously be carried out by engineers, but the lines of professional competence have recently been more closely defined.

To defend their interests, architects have since 1986 organised themselves under the Associação dos Arquitectos Portugueses (AAP). Site supervision and project control are normally carried out by engineers, who also fulfil the role normally undertaken by the quantity surveyor in the UK. Engineering is the dominant profession in the construction industry and engineers also occupy important positions in government and industry and have much influence. There are two types of qualified engineer, the *engenheiros* and the *engenheiros tecnicos*. The former requires a *licenciatura* (a five year university degree qualification) and the latter title is obtained after a less scientifically rigorous three year course.

Contractors in Portugal tend to be smaller than those in Spain and less capable of tackling the largest contracts or of operating outside the country.

Contracts

Government contracts are subject to regulations governing tendering procedures and liability. Consultant's fees are also prescribed. Private activity is less regulated and there is no standard form of contract, though the practice used for public contracts tends to be followed. Occasionally translations of British forms of contract are used. On large contracts the system of separate trades contracting tends to be used, much as in the construction management system used in the UK.

The construction team is usually composed of an architect (who seldom acts as a team leader), a structural engineer, an electrical engineer, a mechanical engineer, a landscape architect and a cost estimator. The construction operation must have a director of works, usually a civil engineer or technical engineer, in addition to an agent representing the lead contractor.

GREECE

Greece has a population of 10 million, very close in size to that of Portugal. Its population density is only 76 people per sq km. Greece and Portugal are the two poorest countries in the Community in terms of GDP per capita, with Greece just ahead of Portugal in 1988. It has fewer cars per head of population than any other Member State. Nevertheless, the air in Athens is highly polluted, and this together with long periods of higher than normal summer temperatures in 1989 and 1990 caused many deaths in the city. Pollution has also ravaged the wealth of historic monuments in the area.

Thirty-eight per cent of the population lives in Athens, the commercial, financial and diplomatic centre of Greece. About half the nation's industry is in Greater Athens, which includes the busy port of Piraeus. The second city is Salonika, a major port with an expanding industry.

Greece joined the European Communities in 1981. It has no borders with other Member States but 60 per cent of its trade is with those countries. The largest volume of imports comes from Germany, followed by Italy, then France and the UK roughly equal. Greece is a major exporter of cement and its accession to the Community has caused problems for the cement industry in some of the other Member States.

Communication in Greece can cause problems. Greek is an intricate language that can take much time to learn. The Greek alphabet, complex grammar and difficult pronunciation all extend the learning period. The existence of dialects is another complication.

The economy

A very high 28.5 per cent of employment is in agriculture, and the industrial sector is uncompetitive. The economy is fragile, suffering from high inflation, low growth and a high level of public sector borrowing. In the five years up to 1987 growth averaged 1.4 per cent and inflation 19.3 per cent, the highest in OECD's lists apart from Yugoslavia. The situation improved slightly after that with inflation dropping to around 13 per cent in 1988/89, and growth increasing to 4 per cent in 1988 before falling back to 2.5 per cent in 1989. However, the public sector borrowing requirement (PSBR) was getting worse, and various income generating measures were introduced, such as a 5 per cent surcharge on income tax for the top 10 per cent of wage earners and a large increase in fuel taxes. One of the problems is severe overmanning in the public sector and an automatic indexing scheme for increasing public sector pay (known as ATA). Incidentally, this was compulsory for the private sector until 1988 and since then the unions have used the ATA rate as the floor level for wage negotiations.

Greek labour, including clerical and professional staff, get paid for

14 months but work only 11 each year. These extra payment months, combined with high social security contributions by the employer, the many additional public holidays and other hidden costs are said to add 70 per cent to the monthly cost of employing individuals. Severance pay obligations strongly discourage the shedding of labour after three to five years of employment.

The general unemployment rate was only 7.8 per cent in 1988, but was expected to grow. Amongst the most highly qualified, however, unemployment is as high as 30 per cent.

The Greek government has obtained loans from the Community amongst other sources and tough conditions have been imposed. These include the substantial cutting back of public sector employment, reducing the scope for tax evasion and reduction in the public sector borrowing requirement.

Construction

Greece accounts for only 0.5 per cent of EC construction. At 4.3 per cent of GDP (in 1988) it is proportionately smaller than in any other Member State. Non-residential output is particularly small, as is the recorded output of repair and maintenance. Just under 6 per cent of the workforce, numbering some 233 000, are officially employed in the construction industry.

Of interest to construction companies are the government's plans to involve private sector funding through concession contracts for infrastructure projects. Bridges and motorways are amongst the projects planned. Like Portugal, Greece is wholly an 'objective 1' region for the purposes of the European structural funds but receives a little less. In 1987 the level of Community grants for Greece was about 85 per cent of that received by Portugal.

Architects and engineers must be registered to practise. Main contractors are also required to enroll on the state register. Fees for design work for public contracts are regulated and published fee scales are the basis of payment for private design work. Most standards used in Greece are adapted from German, British and American versions (Greek professionals have in many cases been educated in the UK or the USA, and English is usually the second language). It is said, however, to be very difficult for foreign firms to carry out work, other than the largest contracts, in Greece because of the complex bye-laws which are often the subject of negotiation. Working with a local partner is naturally the simplest way round this type of problem.

IRELAND

Ireland is the second smallest Member State of the Community in terms of population. Only Luxembourg is smaller. In area, however, it ranks eighth. The population is 3.5 million, a million of whom live in Dublin, the only major city. The population density of 51 per square kilometre is the lowest in the Community. There are twice as many cows as people in Ireland and the agricultural sector, which employs nearly 16 per cent of the workforce (1987 figure), remains crucial. For comparison, in the UK 2.4 per cent were employed in agriculture in 1987, and in western Germany 5.2 per cent. The population is relatively young. The number of 15 to 19 year olds is expected to grow steadily during the 1990s, whereas in the UK the number in the year 2000 is forecast to be 22 per cent down on the 1984 level. In west Germany the position is worse, with a drop of 40 per cent forecast for the same period and the drop in Italy almost as much. Educational attainment tends to be high in Ireland, with the level of spending on education per capita being similar to that in the UK, despite the GDP per capita being about 64 per cent of the UK figure. The high level of unemployment, over 18 per cent at the start of 1989, has resulted in many of the well-qualified young people emigrating, a source of concern to the Irish. In 1988, 32 000 emigrated, which combined with economic recovery reduced the rate of unemployment to 16.7 per cent, a fall of 1 per cent from the previous year.

The proximity to the UK and hundreds of years of colonial dominance, have resulted in close similarities in the institutional and legal frameworks between the two countries. The UK is the major trading partner, purchasing 35 per cent of Irish exports and supplying 42 per cent of Irish imports. Once the UK is linked to the Continent by the Channel Tunnel, Ireland will be the only Member State without a fixed link to other Member States. Its isolation on the western periphery of Europe, and the attendant high cost of transport, is a concern since Ireland is heavily dependent on foreign manufacturers who use it as a base for manufacturing for the European market. A shift of the economic centre of gravity of Europe towards the east is a worry in Dublin since the low labour costs of the eastern and central European countries could undermine Ireland's attractions for foreign industry.

The economy

Gross domestic product per capita and private consumption per capita using current purchasing power parities put Ireland just ahead of Greece and Portugal. The dominant features of the economy are the high levels of government debt, the high rate of unemployment and the large proportion of manufacturing carried out by foreign firms. It has been a member of the European Monetary System (EMS) since 1979. This has required tight macroeconomic policies which have succeeded in bringing inflation and interest rates down in relation to the UK, and almost to German levels. The high rate of unemployment, partly due to tight government policy and partly due to appreciation of the Irish pound in relation to sterling causing a drop of competitiveness in UK markets, has been the cost. In 1986 the problem of the appreciating currency forced an 8 per cent deflation of bands within the EMS.

Ireland has strongly encouraged foreign manufacturers to establish in the country, offering general low corporate tax rates for manufacturing, and tax 'holidays' for foreign firms in their first 10 years of operation. In 1985 foreign firms accounted for 63 per cent of manufacturing output and 41 per cent of manufacturing employment. In 1988 the repatriation of profits had reached a level of 10.5 per cent of GDP. Foreign firms come from Germany, the USA and Japan amongst others and tend to be in 'high-tech' assembly.

Corporate taxes on profits were reduced to 43 per cent in 1989, but manufacturing industry benefits from a rate of 10 per cent. Standard rate VAT is 25 per cent, but the exemptions and lower rated categories of activity bring the effective rate down to 13.4 per cent. Income tax rates in 1989 were 56, 48 and 32 per cent, but again the reliefs and allowances brought the effective rate down to 13 per cent. Reform in the tax system is likely, bringing the rates down from relatively high levels with various distorting exemptions, to levels more in line with Community norms.

The country has benefited financially from membership of the Community. Its large agricultural sector has done well out of the Common Agricultural Policy. Reductions in CAP funding, and the possibility of drastic reform of agricultural policy have caused much concern in Ireland, but this is partly compensated by the increase in European Regional Development Fund grants. Like Greece, Portugal and most of Spain, Ireland is classified as an 'objective 1' area for the purpose of the Community structural funds. Ambitious plans, to make use of Community and private funding, include investment in infrastructure, telecommunications, energy, tourism and agriculture.

The prospects for continued steady growth in the economy appear to be good, though the small market is of limited interest for construction companies.

Construction

In 1988 only 70 000, about 6.5 per cent of total employment, were employed in construction. Employment in the industry had declined in all of the previous five years, though the rate of contraction was slowing with employment in 1988 being 1.2 per cent less than 1987. The trend is likely to be reversed with modest growth in construction due to the improving economy, and increased investment in infrastructure.

The organisation of the professions and the industry, and even the contract procedures used, closely resemble those in the UK. The Institute of Engineers of Ireland, the Royal Institution of Architects of Ireland and the Royal Institution of Chartered Surveyors fulfil functions very similar to those of the Institution of Civil Engineers, the Royal Institute of British Architects and the Royal Institution of Chartered Surveyors of the UK. Nominated sub-contractors and quantity surveyors are also part of the system. The principles of legal liability in both countries are similar, with liabilities in both contract and tort. Many UK companies operate in Ireland, and UK experience is generally a valuable background for setting up construction operations in Ireland.

Construction output as a percentage of GDP is lower in Ireland than in any other Member State. Irish construction output represented a mere 0.7 per cent of the Community total in 1988.

Demand for construction in Ireland is expected to grow during the 1990s, with private non-residential investment being an important source. Increased funding from Community structural funds should improve prospects in the civil engineering sector.

LUXEMBOURG

For motorists, the Grand Duchy of Luxembourg is located somewhere on the motorway between Brussels and Strasbourg. Those driving fast may not notice it. Those more cultivated will know of its wooded beauty. Only 2600 square kilometres in area and inhabited by less than 400 000 people it is similar to Bristol in size. It has borders with Germany, France and Belgium. It is worth a stop, even if only because the petrol is cheap!

The Grand Duchy of Luxembourg has had to remain open to its neighbours. Over a quarter of its population are foreigners and a third of its workforce is made up of cross-border commuters and long-term immigrants. It is also open linguistically and most Luxembourgers are polyglot. Besides a dialect called Letzebutgesch, French and German are the administrative languages and English is widely spoken, particularly in business circles. Luxembourg trades most actively with Germany, while the Luxembourg franc is tied to the Belgian franc. It has no university of its own.

Construction

The construction industry is naturally of limited size and capacity. It therefore plays host to many foreign contractors and consulting firms. These companies find a flexible easy working environment. German and French standards are widely accepted. Work tends to be contracted by separate trades, apart from major projects which are usually undertaken by main contractors.

Construction is regulated not by national law, but by each of the 118 authorities. They set out their own regulations which mainly cover planning and technical requirements. However, in practical terms only Luxembourg City exercises genuine control over construction.

As in Germany, high quality of the workmanship and the incorporated products is crucial. Clients will generally show great concern over the quality of secondary work and finishes.

There is no standard contract for private contracts but specifications vary little from one contract to another. General specifications and award procedures for government contracts are regulated.

Liability and insurance

A 10 year liability system inspired by the *code Napoléon* is used. It parallels aspects of both the Belgian and the French civil code. There is no compulsory insurance and liability for minor work is for two years. Crudely speaking, the liability system in Luxembourg is the French system prior to 1978.

Insurance is not required either by law or by the civil code in Luxembourg. It is taken out either on a voluntary basis or as required by the client in the special specifications. The two types of insurance normally provided give cover for two or 10 year defects liability, and an 'all risks' site insurance policy.

For government works, defects liability insurance is regularly required, except for minor works. The same applies to new housing and large-scale private sector work. This type of insurance does not apply to individual builders, but to the object built. It covers all the parties involved: developers, architects, engineers, contractors and sub-contractors.

The 'all risks' site insurance tends to be required by clients for major work, particularly when undertaken by a general contractor.

Chapter 14

Impacts and Opportunities in the New European Market

With change comes opportunity and there is no shortage of change in Europe. On its own, the single market would have a major impact on the construction industry, but in addition we have the pulling down of the barriers to eastern Europe putting on show a dilapidated and despoiled scene, but one of great promise. We also have spreading alarm about the environmental impact of man's activities, the realisation that unrestricted road traffic is not compatible with a pleasant environment, particularly in towns and cities, and finally there has been another spin of the oil price merry-go-round with war and turmoil in the Middle East. The consequences of some of these changes are vast and impossible to foresee, but what of the internal market, what are the opportunities, what are the pitfalls, and how can consultants and contractors prepare and plan?

The key internal market measures

Standards for products and codes for design are changing, the former only slowly, but the greatest opportunity for influencing their development is when they are still at an early stage of development. The Cassis de Dijon case established the principle that Member States cannot restrict the free movement of goods which are lawfully sold in any Member State. The principle can be extended to construction products, and where there are price differentials, as there are in many instances, these can be exploited. Standards will not only affect the product and its design, but will also affect documentation and procedures, though developments in this area show little sign of progress. Associated with the technical standards is quality assurance which has been pioneered in the UK and which could lead to opportunities in contracting and consulting.

Standards also relate to the environment. Increasingly high standards concerned with reduction in air, water and soil pollution will have direct influence on construction workload, but for those who look beyond the work generated by these changes there is the chance to find direct opportunities, for example in the fields of waste disposal, energy efficiency and energy generation. Consultants may be interested in an existing repercussion of environmental concern – the requirement for 'environmental impact assessments'. Architects have been directed to opportunities in this direction, although without a firm scientific base for their training they would seem not to be fitted to carry out such studies, except in relation to the visual impact of new projects. The impact of projects on urban systems and on the natural environment demands a wide-ranging scientifically orientated background.

Safety standards are likely to become more uniform, and uniformly high. Large contractors in the UK already have high standards in this area but some other Member States have some catching up to do. Safety applies to the construction process, the product itself and the products incorporated therein. Standards are rising and the risks during the transition period when new products and procedures come into use may be heightened. With the tendency towards consumer protection, the reversal of the burden of proof and strict liability, the industry needs to take even greater care. Quality assurance has an important part to play. The UK and some other Member States can offer expertise in this area. Safety plans, which must at an early stage assess risk and plan how to minimise it, offer considerable opportunities.

Greater determination to open up public contracts will have an impact, particularly when the new rules have been implemented in the water, energy, transport and telecommunications sectors, and extended to services. Measures on the procurement of works and supplies have been in existence for some years but without much impact. The improvement is in the measures themselves and also in enforcement. Contractors and consultants will be able to harass authorities who do not comply, either directly or via the Commission, and even to claim damages in some cases. The changes in the rules combined with the requirement to specify by reference to European standards or other common standards, better monitoring and enforcement by the Commission and growing awareness in industry, could make public procurement abroad an important market.

Many new contracts, frequently public works contracts, will be directly related to the development of the single market; the improvement of the road and rail networks with a new emphasis on cross-border integration creates new work in its own right, but the new patterns of movement and trade are stimulating the formation of industrial and commercial nodes of development. Added to this is the large transfer of funds from agriculture into infrastructure, and from the north to the south through the structural funds. Private investment is following the infrastructure. Companies from the north are moving south, and investors are moving in from the USA, Japan and elsewhere.

The Russian economist Kondratieff postulated the existence of a long-wave economic cycle with a period of around 55 years. The last major decline into depression, marked by a stock market bubble and subsequent collapse, occurred in the early 1930s. It was not until the post-war period of reconstruction that the cycle turned and the economies of the West lifted into growth. Looking beyond the immediate period of recession, the prospects for another long-term growth cycle, about 55 years on from the 1950s, stimulated by economic restructuring in the Community and physical reconstruction in the regions around it, are encouraging.

A change in attitude of business

Economists have found it hard to explain economic cycles except in terms of psychology and the conditioning of behaviour, particularly among the business community. The objective of completing the internal market, together with increasing scale of international competition, has been reflected in new patterns of behaviour in industry. Restructuring of industry has taken place; cross-border activity has intensified in terms of collaboration, joint ventures, mergers and takeovers; business has anticipated technical changes which are breaking down the barriers – the barriers that created 'non-Europe'. This could be the sign of fresh vigour in European industry and an end to 'Eurosclerosis'.

Advantages of expanding horizons

Opportunities as well as threats are being introduced into markets, both domestic and international. The internal market initiatives are bringing developments which industry needs to monitor and assess. Brussels is the source of many of the initiatives, but there are also changes taking place in the markets which must be observed. The reactive approach is to respond to the events and initiatives as they take place. The pro-active approach is to anticipate developments, to influence them and to prepare to seize opportunities which will emerge. Companies may choose to remain only in their domestic markets but this decision should only be taken after a thorough assessment of the advantages and risks of expanding into other countries, particularly into the other Member States which are to form the single market. What are the advantages?

Firstly there is the need to grow. Without the ambition and stimulus of growth, passive defensiveness can develop and this may well be a prelude to decline and decay. Facing the challenge of expansion invigorates and growth can bring access to new capital, new skills and new ideas. There is the obvious advantage that increased turnover and size may bring increased profits, but some of the possible benefits are less clear. The linguistic, regulatory, cultural and practical problems involved in working abroad mean that a local partner, or some local recruitment, will be almost indispensible. Such links will themselves enrich the organisations. Each nation brings special talents, practices and technologies. Falling back on national stereotypes, Germans may bring a systematic rigour and insistence on high standards in engineering and thorough planning, the Italians may bring flair and creativity, the French a strategic vision and commitment to technology, the British a pragmatic adaptability, and so on. Those who have made the leap abroad often talk of 'synergy' – where the whole is greater than the sum of the parts. This comes about because an excitement and renewed enthusiasm is generated; the effects can be far-reaching. Expertise in other parts of the Community and beyond brings the capability to offer new and broader services for clients. Companies wanting to invest in buildings and facilities abroad are relieved and grateful to be able to work with those they know, even if the fellow national is merely providing monitoring and communication with the local industry.

Many companies working abroad have made the first step with an existing client. John Laing Management Contracting acknowledged that they went into France 'on the back of' a paper mill project in Strasbourg for an existing client. The Americans often think of the UK as a European base, while the Japanese feel more comfortable dealing with a UK firm for European activities. English is usually the second language for the Japanese and many Japanese companies have investments in the UK.

Companies often get started in overseas markets because of personal preferences and interests of the managers concerned; reasons may include the professional interest in different systems and technologies, the special status that being an expatriate brings, a preferred climate, the cultural richness that many countries enjoy, a move to a less crowded country, advantages for one's children stemming from a preferred educational system and in being bilingual. Simon Conolly of Ian Richie Architects, who has worked in Spain and

France, is pleased to work out of the UK, commenting that the architectural interest was higher, with more cultural work and less emphasis on the commercial. These and many more advantages are cited.

Finally there is the advantage of diversity and spread of risk. The UK and other English speaking countries tend to share a more pronounced business cycle than seen in mainland European countries, and the cycles tend to be aligned. Germany, as a prime example, tends to be more stable and its economic cycles may not parallel those in the UK. The special circumstances surrounding reunification appear to have resulted in a growth in German activity at a time when most European countries have seen some slowdown and the UK has seen recession. Some of the other countries in the Community are at an earlier stage of economic growth than the UK and this results in emphasis on infrastructure and other new construction. Spain, in particular, has seen very rapid development since joining the Community. Its growth rate during 1986–90 averaged 4.5 per cent, in comparison with 3.0 per cent for the UK and 2.6 per cent for west Germany. Construction activity in Spain has shown remarkable growth, which reached 13 per cent in 1989. In 1990, when UK construction activity declined, the Spanish construction output continued to grow, though *only* at 10 per cent.

Demographic differences between Members States may also be important. The UK and Germany have low birth rates and ageing populations. Spain, on the other hand, has a large bulge in the age ranges between 10 and 29, although in the last 10 years its birth rate has fallen and the European pattern is becoming established. Demographic patterns clearly have an impact on the type of construction work required, with new family housing, schools and jobs needed for growing populations of young adults. Sheltered homes, retirement homes and hospitals are needed for countries with increasing populations of middle-aged and elderly.

Factors to consider

Despite the beginnings of convergence there is as yet no single market in construction. There are 12 very different markets. The differences cover building control, great variety in laws and regulations, variation between centralised and localised control and organisation, contract conditions, liability periods and insurance requirements, the number and rigour of standards, safety requirements. There are also differences in the structure of the industry in different countries, roles of the parties, sub-contracting practice, patterns of ownership of the parties and differences in education, training and functions of the people involved.

There are similarities between some Member States in certain respects, but the differences between Member States are so substantial that local people will nearly always be essential. For example, a Belgian site engineer will be accustomed to sites where trades contracting is the norm and would find the greatest difficulty in adapting to the UK environment. In the UK contracts are organised differently and great emphasis is placed on the conditions of contract. The main contractor employs numerous sub-contractors directly, and some of these will be 'nominated' by the architect or engineer. Site organisation, the use of plant and the safety practices will also be rather different.

Until systems and education come more closely into line, transplanting of nationals will remain so difficult that most companies will establish alliances or joint ventures, or acquire or merge with firms in the country where they wish to operate. Mergers and acquisitions were popular in the 1980s, but the industry is moving to a more cautious approach by the establishing of joint ventures and other forms of collaboration. The European Economic Interest Grouping (EEIG) is a form of cross-border entity developed for just this type of collaboration (see chapter 5). Such loose arrangements permit a thorough assessment of a partner's capabilities – a trial marriage before a more permanent bond is made. The risk is reduced, but at the same time the potential return is less.

It is unrealistic for a firm to operate in other Member States without establishing local links, but to obtain the maximum benefit from working abroad some sort of exchange should be promoted. However, most staff in international firms are recruited locally and spend all their working lives in their own country. Exchanges and the impact of a firm's international interests are restricted primarily to senior managers, new graduates, young high-fliers and to scientific and technical specialists. The mobility of key staff results in the assimilation and development of best practice and new ideas. Language and cultural differences can be a problem, although not insuperable even for the British. Companies with an international dimension attract a higher calibre of staff.

A longer-term strategic outlook is needed for the European market than is common in the UK. It can take three years or more to build a useful position in another country. Companies which realised in 1990 that the UK market was shrinking began to have the time and desire to think about the Continental market. But by then a vulnerable position at home discouraged ventures into new markets. The time to have moved would have been three year earlier, but then firms were rushed off their feet and could barely find sufficient staff for domestic commitments. Investment in time and money is needed to explore new markets, find the right partners and to become established.

Special expertise

The diversified companies

Companies which have developed particular technologies or expertise may find new markets easier to penetrate. The general contractor has little to offer other than management expertise and experience, and this may not travel well. The major European contractors are tending to develop a range of specialisms. Bouygues (pronounced *Bwig*) has diversified into television communications. Compagnie Générales des Eaux, another French company, has only 35 per cent of its turnover in construction (through its subsidiary Societé Générale des Eaux). It has interests in water related facilities (24 per cent), thermal and electrical energy (18 per cent) and urban equipment (8 per cent). Many of the major UK construction companies are parts of diversified groups, who for example may have interests in plant hire, mining, quarrying, property development, construction materials, water supply, hotels and house-building. These wider interests may be helpful in getting started abroad.

Construction specialisms

Expertise in specialised areas of construction may be easy to market widely. Underground engineering, tunnelling, deep excavations, piling, pipe crossings of rivers and estuaries are examples. Bachy is a European based company operating in a niche market in piling; Terre Armée (Reinforced Earth Ltd in the UK) has built up a network of companies around Europe offering design and supply services in relation to a patented earth retaining and reinforcing system. Examples abound. Research and development may prove to be a key to European expansion, and firms may find that they can make use of European Community research and development funding to undertake collaborative work. The major UK construction companies generally spend a smaller proportion of their turnover on research and development than construction companies from France and Germany.

Other expertise

Pollution monitoring and control, waste treatment and disposal, treatment of contaminated sites (the German firms Hochtief and Strabag are involved in this), urban transport sytems, energy conservation and energy generation other than by fossil fuels and nuclear means are all specialisms which may have wide potential.

Commercial and management expertise

Increasingly the public sector is trying to harness private sector expertise for infrastructure projects. This is usually by means of concession contracts, known in the UK as 'build, operate and transfer' (BOT) contracts. Commercial experience in these types of contracts is particularly valuable. They have been widely used for tolled motorway and bridge construction and light rail transport systems. Examples include the £500 million contract awarded to John Laing and GTM Entrepose (the French company) for the design and construction of the second bridge across the Severn Estuary in the UK, and for operating both the new and existing bridges, charging tolls; the Dartford crossing, the proposed Skye bridge, Manchester Metrolink light rail systems, car parks, and motorways in France and Italy. Such projects may be part financed by the state, for example most light rail systems being built in the UK can benefit from up to 50 per cent state funding. Transroute and Cofiroute of France (the only 100 per cent private French Autoroute operator), and Italstat of Italy are companies operating in this sector. GTM Entrepose has also won BOT contracts for car park facilities.

Commercial experience is also needed for leisure and tourism projects and for 'business park' developments. For example, a British property group, Arlington Securities, were chosen by Calais Town Council in 1989 to develop a FF 2500 million business park and retail centre on the outskirts of the town. Norwest Holst is linking up with its sister firm AMI, the private health-care giant, to form Britain's first company specialising in the design and construction of hospitals. Bouygues and Kier (now part of the Beazer group) formed a joint venture to construct the Center Parcs holiday village in Elveden Forest, Suffolk. Bouygues provided expertise in the design and construction of the large translucent dome which was needed to exclude the British weather.

Construction management for major projects is another area where expertise can be exported. Bovis acquired expertise in this area by buying a US construction management firm, Lehrer McGovern, and now uses it to win contracts on the Continent as well as to carry out projects in the UK. Lehrer McGovern were the construction management contractor for the Paternoster Square and Canary Wharf developments in London. They are providing construction management for EuroDisneyland near Paris, in Spain and in eastern Europe. In some cases special links with other professionals can be an asset for winning work. Lehrer McGovern had an established relationship with the US architect Cesar Pelli, working with him on skyscraper buildings in New York, and then in London on the Canary Wharf Tower in the docklands project.

Which markets?

Companies need to be selective about which markets and regions to concentrate on when working abroad. Criteria for selection include size, growth potential, stability and the strength of the existing competition. In the Community, Germany has very much the largest construction market, followed by France, the UK, Italy and Spain. Together they account for 80 per cent of Community GDP. One of the advantages of concentrating on the large markets is that volume can be built up, with the least exposure to new systems and languages.

Growth rates are most rapid in Spain, Germany and Portugal. The Spanish economy is still growing at 5 per cent per annum, and 4 per cent per annum is predicted until 1995. Construction output is forecast to maintain 6 per cent per annum growth. Germany has been able to maintain steady economic growth, but the reunification is resulting in a surge of new construction work in east Germany. It is estimated that £100 billion will be needed to upgrade the infrastructure in that part of Germany to west German standards. Industrial investment is also building up, though not fast enough to cope with the loss of jobs from the shutting down of old-fashioned, heavy and polluting industry. Another source of work is the reinstatement of contaminated land.

The four largest Community Member States all have well-established domestic construction industries which can in normal circumstances meet domestic needs. Rapid development in some of the poorer regions of the Community, which is being stimulated by the European Regional Development Funds, is leading to a demand for construction which the local industries are not capable of meeting. The considerable funds, which are intended to be supplementary to national funding, are concentrated on infrastructure development and are being targeted on Greece, Ireland (north and south), most of Spain, Portugal and the south of Italy. One other reason for following the structural funds is that infrastructure contracts supported by the funds must be let in accordance with the public procurement rules (see chapter 6). These are designed to ensure that the contracts are open to foreign competition. Also the contracts are advertised in the *Official Journal* and in *Tenders Electronic Daily* (TED). Associated with the infrastructure work there is the knock-on effect in other sectors which benefit from increased national spending and income. Not only do national firms respond by increasing investment, but foreign firms are attracted by the growth prospects and also invest in new facilities. These parts of the Community are those with low wage rates, which also attract foreign investment. Many firms from the north of Europe, from Japan, the USA and elsewhere are setting up in these countries, particularly Spain, Portugal and Ireland.

Communication routes have always led to settlement and development. Cross-border trade in the Community is expanding. The Member States are developing a European transport network, in some cases supported by Community funds. Nodes in this network are natural growth points for industry and population. Apart from the infrastructure construction itself, private industry is anxious to invest around the transport axes which will lead to new corridors of development. Parts of France, southern Germany, northern Spain and Italy are likely to be beneficiaries. One of the problem areas in European transport routes is the neck of Austria which lies between Italy and Germany. The Austrians are very unhappy about existing levels of heavy goods vehicles causing atmospheric and noise pollution. There is pressure to shift goods and passengers to the railways and to minimise the impact of such railways by the use of tunnels, cuttings and earth mounds wherever possible. East-west transport routes will also need upgrading now that eastern Europe is becoming reintegrated with the west. East-west trade has been very limited, partly because of political and military problems and partly because of the small size and backwardness of the eastern economies. As the east develops, trade and tourism will increase. This creates a demand for better communications and new infrastructure. Again private investment along routes and at nodes is attracted.

Trends in Community construction markets

Private sector projects

Clients for large private sector projects are demanding more speed in design and construction from the construction industry. Projects are often divided into packages or carried out on a turnkey basis by design and build contractors. Installations are becoming more complex. Contractors for the major projects need financial strength and technological competence and the less capable firms will find themselves working as sub-contractors and marginalised into smaller and more basic contracts. Multinational companies which can offer a full range of services from feasibility studies through construction to commissioning and maintenance have the advantage, particularly if they have experience with sophisticated products and technologies.

Clients are turning towards contractors who can offer wide-ranging 'turnkey' services. Clients are also becoming keen to develop privileged relationships with contractors, in what the Americans call 'team working'. A contractor who gets to know his client's business and can offer a complete service from project appraisal to commissioning of a new facility will be at a great advantage. The single market is bringing new groupings into being in industries such as food, manufacturing, banking and insurance. Investment in other Member States and the introduction of new technology in shorter timescales puts pressure on these companies. A contractor who can offer a package covering land acquisition, planning, financing, organisation and production technology, and possibly even recruitment and training, is very valuable. The trend to turnkey contracts may tend to displace the smaller consultants and contractors. Professionals who remain organised into small practices will not be able to offer the breadth of skills required by the multinational client or keep up with the latest developments.

The public sector

The scale of public sector projects is unlikely to change much. Small and medium-scale work will continue to be carried out by local firms, or large firms with a local presence.

Protectionist tendencies evident in many regions may survive the new regime in public procurement legislation, but mainly in connection with smaller projects below the thresholds. The advance notices for public projects which contracting authorities must publish in the *Official Journal* (reproduced in *Tenders Electronic Daily*) will assist foreign contractors to prepare themselves for competition. Rapid growth in some countries stimulated by regional development funding is leading to much new civil engineering and public work. The Public Works Directive, tougher Commission policing and the compliance directive should ensure open competition.

Concession, or build-operate-transfer (BOT) contracts (see p 250) are becoming more widely used. The public sector is preferring to hand over responsibility for design, construction and operation to the private sector using such contracts, or through privatisation. The privatisation of non-nuclear power, the telecommunications and water industries in the UK are examples. However, from 1 January 1993 these industries, even if privatised, will be subject to rules similar to those governing public works contracts (1 January 1996 in Spain and 1 January 1998 for Spain and Portugal). Considerable volumes of work in these sectors are likely. In the Community water industry investment of £75 billion over the next five years is forecast, and in the UK £30 billion over the next 10 years.

Among activities being returned to the private sector is likely to be in-house design and consultancy. This is a development which the Commission is seeking to promote since work carried out in-house is clearly excluded from the internal market and probably costs the public rather more than it need. Greece is one country where the reduction of in-house design is likely to take place soon since the Community has required the Greek government to cut back public sector employment as a loan condition. In-house design of public work is prevalent in Greece.

Industry responses

The largest European companies with construction interests have been building up their positions in the European market. Companies such as SAE of France, Philip Holzmann of Germany, AMEC of the UK, Dragados y Construcciones of Spain, appear to have well-formulated strategies in operation. Growing markets are resulting in mergers and acquisitions, as well as other forms of collaboration. A recent example is the merger between two French giants, Lyonnais des Eaux and Dumez. The powerful new entity has a wide range of skills from water supply and treatment, through energy, waste collection, electrical distribution, building products to developed design and construction skills in building and civil engineering. AMEC of the UK was formed in 1982 after the merger of Fairclough and William Press. The company went on to absorb Matthew Hall, a large mechanical and process engineering firm. Many companies believe that size is needed to survive after 1992.

As well as building up positions in the internal market, these large groupings are starting to compete in eastern European markets and further afield against the Japanese and US majors, some of which are very large. For example, Taisei of Japan has twice the turnover of Tarmac. Three US firms stood at the top of the *Engineering News Record* league table in 1988. Fluor Daniel Inc was top with 1988 contracts worth $12.2 billion, Bechtel was second with 1988 contracts worth $10.9 billion and Kellog was third with 1988 contracts worth $9.3 billion. As well as expertise in building and civil engineering, the top US companies tend also to offer capability in energy, process engineering, offshore and waste treatment and disposal. An *Engineering News Record* survey in 1990 of cross-border orders won by the world's largest contractors showed the European export market to be the largest and fastest growing. In 1989, orders awarded to foreign firms in European countries were worth $25 billion. The equivalent figures for Asia and the Middle East were $24 billion and $18 billion respectively, with the Middle East market showing virtually no increase over the previous year.

Surprisingly, even the large companies find difficulty in keeping abreast of new initiatives and legislation emanating from Brussels and tend to rely on consultants to keep them informed on relevant developments.

For some years the medium-sized companies (100 to 500 employees) have tended to merge or be absorbed by large companies. Over the last 15 years in France, the large national firms have gained at the expense of the medium-sized regional firms. In the smaller and southern Member States, however, the medium-sized companies still form the backbone of the national industries.

Medium-sized companies may be of interest for companies seeking to become established in a new market. The local company may welcome a link with another firm to strengthen its position locally through access to funds and expertise. Companies in the smaller Member States, or near to borders, such as in the Alsace region of France, can be valuable in that they are often cosmopolitan, cross-border orientated and have had to develop language and expertise in technical and regulatory aspects. Even small firms located near borders are often comfortable working 'abroad'. Ominium Technique Européen, a small design office based in Strasbourg in the northeast corner of France, undertakes work in Germany and Luxembourg as well as in France.

Single market measures will affect all companies, but small companies with a secure local reputation and client base are unlikely to be threatened by developments which preoccupy the larger firms. Small firms should be able to offer good service and low overheads. Clients also tend to like to work with firms they have dealt with before and know well. Small firms, employing local labour, can also offer rapid response when problems occur. The multinational firms do not threaten such firms. Nevertheless, the small firms may find themselves working with the multinationals as sub-contractors and may also be exposed to new materials, methods and contractual arrangements and liabilities.

Conclusion

The European construction market remains fragmented. Attempts to bring harmonisation will be slow and may barely be noticed by the smaller companies. The larger companies are gathering resources and expertise in preparation for more demanding and sophisticated projects for larger and more international clients. Those willing to face the challenges will be able to enjoy interest and opportunities in the new single market.

Sources and Further Reading

This book aims to provide information about current developments in the European Community and the Community and national contexts. It does not aim to be comprehensive and in any case such information dates rapidly. Most readers concerned with EC matters or wishing to develop links or business in other Member States will need to delve further and get to grips with the various sources of information. Problems can arise from the sheer bulk of information available, the difficulty identifying what is needed, the widespread use of jargon and acronyms, and unfamiliarity with the organisations, systems and concepts involved. Also, market reports and directories are often costly and soon out of date.

The following pages provide a personal and selective guide to the sources of information available both about EC legislation and programmes and about specific Member States. Some suggestions for further reading are also made. Where sources are specifically national they are quoted for the UK but similar facilities exist in other Member States.

Coverage of European matters in the daily press varies widely. The *Financial Times* is probably the most comprehensive on EC matters and is the most widely read daily in the Community and in Brussels. Most key issues are picked up and there are often specialist articles on EC matters, European industries, Member States and regions. The periodic surveys of countries, regions, industrial sectors and other matters are usually valuable.

In the UK the Department of Trade and Industry (DTI) is the essential starting point for current information. The DTI provides information packages about Community developments and institutions, and specific information about other countries and markets. Since 1988 the Department has been promoting the 1992 campaign to raise awareness of the single market and EC opportunities. Related to construction, the Department of the Environment publishes the excellent *Euronews Construction* (see below for details). The British Standards Institute also helps with an 'information for exporters' section.

Official documents of the European Community and much other information related to EC matters is provided in the UK by Her Majesty's Stationery Office (HMSO). The problem is how to track down what one needs so that a specific order can be placed. Incidentally HMSO now provides a fax service for documents or parts of them (see below for details).

The network of European Documentation Centres (CDEs) are invaluable since most libraries do not stock the *Official Journal* or other Community publications (see later for details).

Anyone seriously interested in research, in keeping up with developments or in tendering opportunities will need to use one or more of the available electronic databases provided by the Commission and others. Apart from the ability to obtain up to date information without subscribing to expensive and voluminous journals, periodicals and directories all databases have powerful searching facilities. Some databases provide full text while others provide only bibliographic details. In the case of EC tender notices the use of TED (*Tenders Electronic Daily*) also gives a significant time advantage for tender preparation.

One benefit of using Euro Info Centres (EICs) is that they have free access to certain databases (TED for example). Otherwise costs can be high. The Euro Info Centres also have staff skilled in conducting searches. The practicalities of conducting searches oneself can be a barrier, particularly for computer-phobics. The databases listed later are a selection from those available.

Note: UK telephone and facsimile (fax) numbers are as dialled from the UK. From outside the UK the initial 0 should be replaced by the appropriate international dialling code and the UK country code (44). Numbers outside the UK include the country code (e.g. 32 in Belgium). The + indicates the appropriate international dialling code (e.g. 00 in Belgium). Thus the full telephone number of the Commission in Brussels when calling from the UK is 010 32 2 235 11 11.

UK Government sources

Department of Trade and Industry

The DTI's Single Market Hotline (tel. 081-200 1992) is an excellent starting point for any study on current European Community developments. The following clear and sometimes substantial booklets are available via the Hotline: *The Single Market – The Facts, Action Checklist for Business, Single Market News* (quarterly), *Action guide for smaller firms '1992-for you', Guide to practical advice for business* (national and regional lists), *'Brussels can you hear me?' – How to influence the decision-making process, Europe open for professions, Guide to public purchasing, Financial services, Company law harmonisation, Progress on the Commission White Paper* (updated regularly) and a variety of booklets about European standards in product areas such as construction products and mobile machinery and lifting equipment.

Three short videos are also available. DTI export publications are available in summary form from DTI Export Publications, PO Box 55, Stratford-on-Avon CV37 9GE. For the reports telephone (0789) 296212.

For information about EC Structural Funds contact the World Aid Section of the DTI, Room 042, 1-19 Victoria Street, London SW1H 0ET, tel: 071-215 4255.

Department of the Environment (DoE)

Euronews Construction – this is a very useful quarterly news-sheet produced by the Department of the Environment, Construction Directorate. It is bound with periodic issues of *Building*, the well-known UK periodical but can be obtained direct (tel: 071-276 3146). It gives the inside picture on the development of legislation and documents related to construction, such as the interpretative documents associated with the Construction Products Directive and Eurocodes, Helpfully, articles on most topics give a contact name and telephone number for further information.

Health and Safety Executive

The HSE has information about European legislation under development in its sector, for which it organises the UK consultation during the legislative process.

Contact: HSE, Baynard House, 1 Chepstow Place, London W2 4TF, tel: 071-229 3456.

UK Permanent Representative in Brussels

'UK-rep' is useful for establishing the latest situation relating to the legislative process and in connection with lobbying or the submission of applications to the Community institutions for employment or grants.

Contact: Office of the United Kingdom Permanent Representative to the European Communities, rond-point Schuman 6, B-1040 Bruxelles, tel: Brussels (+32 2) 230 6205.

Commission of the European Communities and associated organisations

Commission of the European Communities

rue de la Loi, B-1049 Bruxelles, Belgium, tel: Brussels (+32 2) 235 11 11. This is the general address for all Commission business. If possible write to the appropriate part of the Commission or the individual with specific responsibility (see below for information on responsibilities within the Commission).

The Commission publishes the *Directory of Community legislation in force and other acts of the Community institutions*. It is updated twice a year and covers secondary legislation such as regulations, decisions, directives, other legislation and agreements between the EC and other countries. The directory includes a chronological index of documents and an index of keywords. *Price:* ECU 75; *Catalogue number:* FX 56 90 000 ENC (available in the UK from HMSO – see details below).

Commission of the European Communities, UK office

8 Storey's Gate, London SW1P 3AU, tel: 071-222 8122.

Eurostat (information office)

Eurostat publishes statistical information about the Community and Member States. L-2920, Luxembourg, tel: (+352) 4301 4567.

Euro Info Centres (EICs)

EICs are set up and organised by the Commission, DGXXIII Small and Medium Enterprises (SME) task force in Brussels, which can be contacted at 80, rue d'Arlon, 1040 Bruxelles, tel: Brussels (+32 2) 235 37 39; fax: (+32 2) 235 73 35.

Euro Info Centres can be very useful sources of information about EC developments and contacts in other Member States for business or academic purposes. See chapter 5 for further information about Euro Info Centres.

The Department of Trade and Industry provides a full list of the UK Euro Info Centres in *Guide to practical advice for business – national list* available on the DTI Hotline 081-200 1992. The centres include the Thames Valley Euro Info Centre, 2–6 Bath Road, Slough, Berks SL1 3SB, tel: (0753) 77877; fax: (0753) 24644 and other centres in Belfast, Brighton, Bristol, Exeter, Inverness, Leeds, Bradford, Leicester, Liverpool, Maidstone, Manchester, Norwich, Nottingham, Pontyprydd, Telford, Stafford and Winchester.

European Documentation Centres (EDCs)

The Community maintains a number of European Community Documentation Centres primarily for educational purposes, though under the terms of the contracts of establishment they should be accessible to the public. They are usually in university libraries in the Member States and are supplied with all official Community Documents. The national office of the Commission of the EC can provide details of all the centres. There are over 60 EDCs in the UK. One such is at the British Library, Great Russell Street, London, tel: 071-636 1544. The Bodleian library of Oxford University and the library of the University of Reading are also EDCs.

Electronic Databases

Note that access to electronic databases requires a suitably equipped microcomputer and a password to the national data network. For information on the latter in the UK contact:

Mr Ken Wilson, National Specialist Sales, 8th Floor, Tenter House, 45 Moorfields, London EC2Y 9TH, tel: 071-250 8719; fax: 071-250 8343.

CELEX

CELEX is the official database on Community law and is widely used by Commission staff and lawyers. The search system is unforgiving but CELEX contains information shortly after it has been published in the *Official Journal*. It contains the treaties and latest modifications (such as the Single European Act), legislation such as directives and regulations, reports of cases before the Court of Justice, preparatory documents such as Commission proposals, resolutions of the European Parliament, opinions of the Economic and Social Committee and parliamentary questions and answers. In most cases the full texts of documents are on the database though lack of capitalisation or the use of upper-case throughout, and the lact of normal formatting impedes reading.

CELEX costs ECU 96 per hour (about £1 per minute). Details from Commission of the European Communities, Eurobases Section (DG IX-E-3), rue de la Loi 200, B-1049 Bruxelles, tel: Brussels (+32 2) 235 00 01/02/03 or (+32 2) 235 90 07 (Mrs Isabelle Dujardin-Kardacz); fax: (+32 2) 236 0740).

SCAD – the Community Documentation Access System

SCAD is valuable for obtaining up to date bibliographic information on Community developments and documentation and is a quick means of tracing information. Having traced a document CELEX can be used for obtaining the full text. SCAD also contains the INFO 92 database which has information about the internal market programme. SCAD has abstracts on EC related topics obtained from scanning 1500 or more periodicals worldwide.

SCAD costs ECU48 per hour while INFO 92 costs ECU 10 per hour. Use the same address as for CELEX for obtaining information about SCAD and INFO 92.

TED (*Tenders Electronic Daily*)

This is updated daily and contains full notices about contracts being offered for tender in more than 80 countries. The notices are as published in the S series of the *Official Journal* and are displayed in TED on the morning of their publication. It is also possible to obtain TED notices corresponding to selected criteria (e.g. civil engineering in Spain) by telex.

TED costs ECU 36 per hour. Information from ECHO Customer Service, BP 2373, L-1023 Luxembourg, tel: (+352) 38 80 41; fax: (+352) 48 80 40.

JUSTIS

JUSTIS is a UK database containing CELEX, information about progress on completing the internal market and UK case reports. The search system is said to be easier to use than that on CELEX itself.

It costs £120 + VAT per month, with no further charges, to use.

Details available from Context Legal Systems Ltd. Tranley House, Tranley Mews, London NW3 2QW, tel: 071-267 7055; fax: 071-267 2745.

JUSTIS is also available on CD-ROM (compact disc – read only memory). Amazingly, the full database, including CELEX, fits on a single CD (compact disc) which is readable on a suitably equipped microcomputer. The initial cost is high and the disc is never fully up to date (they are replaced every six months), but there is not the pressure of high usage cost. As a library resource CD-ROM, and the JUSTIS CD-ROM in particular, is superb.

Contact: Context Legal Systems Ltd as for JUSTIS above.

SPEARHEAD

SPEARHEAD includes a wide range of information about EC and other matters. Apart from single market information provided by the DTI it also contains abstract information from a wide range of publications. It is powerful but costly to register and use. Details from Profile Information, Sunbury House, 79 Staines Road West, Sunbury-on-Thames, Middlesex TW16 7AH, tel: (0932) 76144.

Market information

NEDO and Euroconstruct

The authoritative source of data on the construction industry in Europe (not just the EC) is Euroconstruct. The British representative on Euroconstruct is the National Economic Development Office (NEDO), Millbank Tower, Millbank, London SW1P 4QX, tel: 071-217 4000.

Euroconstruct surveys are produced bi-annually, cost £200 and include economic data and expert analysis – around five to 18 pages per country. Surveys include country and sector forecasts and occasional articles on specialised topics such as civil engineering.

NEDO has also published data on the construction industry in the EC in *Gateway to Europe*, ISBN 07292 09717 published in 1990 at £80 and available from NEDO: tel: 071-217 4037.

Association of Consulting Engineers (ACE), Foreningen af Rådgivende Ingeniører (FRI) and Orde van Netherlandse Raadevende Ingenieurs (ONRI)

These organisations have jointly published reports on engineering consultancy in the European Community. The full set of 11 reports (986 pages in all) costs £600 or they can be purchased individually for £80 each. The reports cover UK, Ireland, Belgium and Luxembourg, the Netherlands, Italy, Germany, Spain, Portugal, France, Greece, Denmark, and are published by Thomas Telford, London (1990).

National Council of Building Materials Producers

The National Council of Building Materials Producers publishes comprehensive reports on France and Germany at £50 each. For information telephone 071-323 3770.

The Royal Institute of British Architects (RIBA)

RIBA is publishing a set of guides on architectural practice in EC Member States. The guide to France, compiled by Davis Langdon and Everest Consultancy Group, is priced at £95 for RIBA members and £150 for non-members. The guides to Spain, Portugal and Germany cost £125 (RIBA members) or £165 (non-members) each and that to Italy is in preparation (1991).

Contact Ian Pritchard of RIBA on 071-580 5533 or RIBA Publications Ltd at Finsbury Mission, 39 Moreland Street, London EC1V 8BB, tel: 071-251 0791; fax: 071-608 2375.

Construction Industry Research and Information Association (CIRIA)

CIRIA publishes guides for UK professionals on the French construction industry (by J. L. Meikle and P. M. Hillebrandt); the West German construction industry (by S. D. Biggs, M. Betts, and M. J. Cottle); the Iberian construction industry (by S. D. Reynolds and S. Sheppard) each priced at £150 (or £75 for CIRIA members).

Contact: CIRIA, 6 Storey's Gate, Westminster, London SW1P 3AU, tel: 071-221 8891.

Updating services

Building Europe (a periodical produced by *Building* magazine) provides an updating service on EC legislation, European construction markets and articles on the influence of the harmonisation of the European market on competition and work practices.

It costs £100 per annum and is available from Builder House, 1 Millharbour, London E14 9RA, tel: 071-737 2222; fax: 071-537 2007.

Suppliers of EC documentation in the UK

HMSO

HMSO are suppliers of government and EC documents and agents for other organisations such as the Organisation for Economic Cooperation and Development (OECD).

Contact: Publicity Department, St Crispin's, Duke Street, Norwich NR3 1PD, or for EC publications enquiries telephone 071-873 8372; fax: 071-873 8463. Orders should go to 511 Nine Elms Lane, London SW8 5DR, tel: 071-873 9090 (quote the title and ISBN/ISSN number). Telephone and credit card orders can be placed on 071-873 9090.

There is also a fax transmission service for *Official Journal* L, C and S Series and for COMdocs. For information enquire about SCANFAX tel: 071-873 8220 or fax: 071-873 8416. Subscription to the L and C series of the *Official Journal* costs £294 and for the S series (supplement – contains tender notices) £204.

Suggestions for further reading and reference

(1) Florence Nicholas (1988) *Common Standards for Enterprises*, Commission of the European Communities, ISBN 92 825 8554 9, £6.75.

(2) Jean-Claude Séché (1988) *A Guide to Working in a Europe without Frontiers*, Commission of the European Communities, ISBN 92 825 8067 9, £14.25.

(3) *Research on the 'cost of non-Europe'* (1988) Basic Findings Volume 1, Commission of the European Communities ISBN 92 825 8605 7.

(4) Commission of the European Communities (1991) *XXIVth general report on the activities of the European Communities*, Brussels, ISBN 92 826 2176 6. This contains information about plans for legislation and progress in most sectors.

(5) Commission of the European Communities (1989) *Guide to the reform of the Community's Structural Funds,* Office for Official Publications of the European Communities, Luxembourg, ISBN 92 826 0029 7.
(6) Clifford Chance (1990) *CCH Guide to 1993, Changes in EEC Law,* 2nd ed, CCH Editions Ltd.
(7) P. A. Cox (ed.), (Summer 1991) *Civil Engineering Project Procedure in the EC,* Thomas Telford, London, UK £35.00.
(8) *Dod's European Companion 1991,* £105. This is a guide to prominent EC individuals and institutions and is available from DOD, 60 Chandos Place, London WC2N 4HG, tel: 071-340 3902.
(9) Richard Owen and Michael Dynes (1989) *The Times Guide to 1992.*
(10) Richard J. Grover & Anton Lavers, *Impact of European Community Policy on Quality Management in Construction,* CIRIA, (expected publication date September 1991).
(11) John Mole, *Mind Your Manners: Culture Clash in the European Single Market* (1990) Industrial Press, £14.95.
(12) Prof. Gerhard Drees (Institute for Construction Management, University of Stuttgart), 'Contractual conditions for construction projects in Germany', *Professional Builder* (South Africa), 4 July 1986, pp 22–3.
(13) J. Knocke (ed.), (expected publication date December 1991). *Post Construction Liability,* E & F N Spon.
(14) David Medhurst (1990) *A Brief and Practical Guide to EC Law,* Blackwell Scientific Publications, Oxford.
(15) *Yearbook of the European Communities and of the other European organisations,* Editions Delta, 55 rue Scailquin, B-1030, Bruxelles, tel: Brussels (+32 2) 217 55 55; fax: Brussels (+32 2) 217 93 93.
(16) *1990 Directory of EEC Information Sources,* Euroconfidential 1990, rue de Rixensart 18, 1332 Genval, Belgium; costs Belgian Francs 7700 (about £120).
(17) *The European 1992 Directory, research and information guide* 2nd edn (1990) ITCU/Coventry Polytechnic, HMSO, ISBN 011 701491 5, cost £12.95.
(18) *Directory of the European Construction Industry 1991,* cost: £85, Available from: Building Economics Bureau Ltd, 5 Carlton Chambers, Shortlands, Bromley, Kent BR2 0EY, tel: 081-464 4518; fax: 081-313 3363.
(19) Iain Wishart, 'Building in the European Communities', *Construction Law Journal,* 1990, Volume 6 pp 3–6.
(20) *European Municipal Directory,* cost: £98, Available from European Directories Ltd, Sellarsbrooke Park, Ganarew, Monmouth,

Gwent NP5 3SS, UK, tel: (0600) 890506; fax: (0600) 890976. Apart from information about local authorities and how local government throughout the Community works the directory contains information about tendering procedures (public procurement), standards, finance budgets, recruitment, etc.

(21) Organisation for Economic Co-operation and Development (OECD) publications, particularly the country studies, are an excellent source of information about economic matters. The HMSO (see above) is the UK agent for OECD publications.

Organisations

CEN/CENELEC (the Committees for European Standardisation, and European Electrotechnical Standardisation)

rue Brederode 2, Box 2, B-1000 Bruxelles, Belgium, tel: CEN Brussels (+32 2) 519 68811; CENELEC (+32 2) 519 6871.

FIEC (International European Construction Federation)

3, rue de Berri, F-75008 Paris, tel: (+33 1) 45 63 11 44; telex: FNTP 640675; fax: (+33 1) 45 61 04 47.

CEBI (Comité Européen des Bureaux d'Ingéniérie/European Committee of Consulting Firms)
and
CEDIC (Comité Européen des Ingénieurs-Conseils du Marché Commun/European Committee of the Consulting Engineers of the Common Market

Avenue Louise 430, Box 5, B-1050 Bruxelles, Belgium, tel: (+32 2) 648 0695; telex: 21591 CEBI; fax: (32 2) 646 0541.

CLAEU (Comité de Liaison des Architectes de l'Europe Unie)

rue de Livourne 158, Box 5, B-1050 Bruxelles, tel: (+32 2) 647 0669.

Institution of Civil Engineers (ICE)

1–7 Great George Street, London SW1P 3AA, tel: 071-222 7722. The

ICE, in common with the other professional institutions, has an officer with responsibility for European matters and a library with relevant publications. The library operates an electronic bulletin board, a branch of which has information about directives, standards, European organisations, research and development funding and other matters connected with Europe.

British Standards Institution (BSI)

BSI (sales, information and library), Linford Wood, Milton Keynes, MK14 6LE, tel: 0908 221166; fax: 0908 320856.

Jobs in Europe

Overseas Placing Unit of the Employment Service, OPS, 4th Floor, Steel City House, Sheffield S1 4PQ.

References

Chapter 1

1.1 COM(85) 310 available from HMSO. ISBN: 92 825 5436 8.

1.2 OJ L169, 29 June 1987; see also reference 4.1.

1.3 89/106/EEC OJ L40/12 11 February 1989.

1.4 KPMG data, quoted in the *Financial Times* 28 January 1991.

1.5 *How European Are We?*, Paragon Communications, 142 Wardour Street, London W1V 3AU, tel: 071-734 6030.

1.6 TED is accessed on-line via any suitably equipped personal computer and packet switched line. Details from ECHC Customer Service, BP 2373, L-1023 Luxembourg, tel: (from UK 010+) 352 48 80 41. Alternatively contact the nearest Euro Info Centre (see later).

Chapter 4

4.1 Office for Official Publications of the European Communities, Treaties establishing the European Communities (abridged edition), Luxembourg 1987 (ISBN: 92 825 7657 4).

4.2 Case 45/87 *Commission of the European Communities* v *Ireland* 22/9/88 (*see reference 6.2*).

4.3 Council Regulation 2137/85 OJ L199 31.8.85.

4.4 Council regulation (EEC) 17 of 6 February 1962, OJ 1962. p204, Special edition 1959–60 p. 87.

4.5 Council Directive 71/305/EEC, OJ L185 15.8.81 p.5, Sp edn 1971 (II) p.682.

4.6 Council Directive 85/374/EEC, OJ L210 7.8.85 p. 29.

4.7 Case 148/78 *Pubblico Ministero* v. *Ratti* [1979] ECR 1629, [1980] CMLR 96, (Court of Justice 1979).

4.8 Case 152/84 *Marshall* v. *Southampton and South West Hampshire Area Health Authority* [1986 ECR 723, [1986] 1 CMLR 688, (European Court of Justice 1986).

4.9 Commission Decision 86/138/EEC Eurofix/Hilti, OJ L65 11.3.88 p. 19.

4.10 Commission Decision 88/568/EEC, OJ L311 17/11/88 p.36.

4.11 Council Directive of 25 July 1985 on the 'approximation of the laws, regulations and administrative provisions of the Member States concerning liability for defective products' 85/374/EEC, OJ L210 7 August 1985, p.29.

4.12 Council Directive 89/440/EEC of 18 July 1989, OJ 21 July 1989 L210 page 1.

4.13 Council Directive 71/305/EEC (as 4.5).

4.14 *Litster* v. *Forth Dry Dock Co Ltd* (1989) 2 WLR 634, [1989] 1 All ER 1134.

4.15 Council Directive 77/187/EEC, OJ L61 5.3.77 p.26.

4.16 Joined cases 271/83, 15/84 and others, *Alan Ainsworth and others* v. *Commission and Council of the European Communities* ECR [1987] pp.167 to 222. (European Court of Justice 1987).

4.17 Case 36/74 *Walrave and Koch* v. *Association Union Cycliste Internationale* [1974] ECR 1405, [1975] CMLR 320 (European Court of Justice 1974).

4.18 Case 43/87 *Defrenne* v. *SA Belge de Navigation Aérienne Sabena* [1976]ECR 455, [1976] 2CMLR 98 (European Court of Justice 1976).

4.19 Case 120/86 *Mulder* v. *Minister van Landbouw en Visserij* [1989] 2CMLR 1 (European Court of Justice 1989).

4.20 Case 4/73 *Nold KG* v. *Commission* [1974] ECR 491, [1974] 2 CMLR 338 (European Court of Justice 1989).

4.21 Case 145/83 & 53/84 *Adams* v. *European Community Commission* [1985] ECR 3539, [1986] 1 CMLR 506 (European Court of Justice 1984).

4.22 *Bourgoin SA* v. *Ministry of Agriculture* [1986] QB 716, [1986] 1 CMLR 267.

4.23 *R* v. *Secretary of State for Transport ex Parte Factortame Ltd* [1989] 3 CMLR, 1 [1989] 2 WLR 997.

4.23 Case 8/74 *Procureur du Roi* v. *Dassonville* [1974]ECR 837, [1974] 2 CMLR (European Court of Justice 1974).

4.24 Case 120/78 *Rewe-Zentral AG* v. *Bundesmonopolverwaltung für Branntwein* [1979] ECR 649, [1979] 3 CMLR 494 (the Cassis de Dijon case).

4.25 *Rush Portuguesa Limitada* v. *Office National d'Immigration, The Times* 12 April 1990 (European Court of Justice 1990).

4.26 Council Directive 64/221/EEC (OJ 56 4.4.64 p 850, Sp edn 1963-64 p1150).

Chapter 5

5.1 Case 8/72 *Cementhandelharen* v. *Commission* [1972] ECR 977, [1973] CMLR 7.

5.2 Case 258/78 *Nungesser* v. *Commission* [1982] ECR 2015, [1983] 1 CMLR 278.

5.3 Council Regulation 2423/88, 11 July 1988, OJ L209 2 August 1988 p.1.

5.4 Commission Regulation No. 4064/89, OJ L395, 1989 p.1.

5.5 Twelfth Directive on single member private limited companies, OJ L395, 30/12/89, pp.40–42.

5.6 Proposal for a European company statute, OJ C263, 16/10/89, pp.41–72 and Bulletin of the EC, Supplement 5/89, £3.00, ISBN 92 826 0931.

5.7 Council Regulation, No. 2137/85 OJ L199 1985 p.1.

Chapter 6

6.1 As 4.25 above.
6.2 Case 45/87 Dundalk Water Supply System, Re: *EC Commission* v. *Ireland* [1989] 1 CMLR 225, [1987] ECR 1369.
6.3 COM(90) 287 – SYN 292, OJ C216, 31 August 1990.
6.4 Committee for Building and Civil Engineering Economic Development, The Public Client and the Construction Industries, National Economic Development Office, HMSO 1975, ISBN: 011 700 739 0*.
6.5 Case 21/88 *Du Pont de Nemours Italiana SpA* v. *Unita Sanitaria Locale No. 2 di Carrara*, The Times Law Reports, 11 April 1990.
6.6 Council Directive 71/304/EEC, OJ L185 of 16 July 1971, p.1.
6.7 Council Directive 71/305/EEC, OJ L185 of 16 July 1971, p.2.
6.8 Council Declaration OJ C82/13 16 August 1971 p.55.
6.9 Council Directive 77/62/EEC, OJ L13/1 15 January 1977.
6.10 Council Directive 80/767/EEC, OJ L215/1 18 August 1980.
6.11 Case 31/87 *Gebroeders Beentjes BV* v. *Netherlands* [1990] 1 CMLR 287.
6.12 Council Directive 89/440/EEC, OJ L210/1 21 July 1989.
6.13 Joined cases including 27/86 *Constructions et Entreprises Industrielles* v. *Association Intercommunale pour les Autoroutes des Ardennes*, case 28/86 *A. Bellini and Co. SpA* v. *Regie de Batiments* and case 29/86 *A. Bellini and Co. SpA* v. *Belgian state*.
6.14 as 6.11 above.
6.15 Proposal for a directive on the procurement procedures of entities providing water, energy, and transport services COM(88) 377 final, 11 October 1988, OJ C264 16 October 1989 p.22.
6.16 A proposal for a directive on the procurement procedures of entities operating in the telecommunications sector COM(88) 378 final, 11 October 1988, OJ C40 17 February 1989.
6.17 Commission of the European Communities, Communication on a Community regime for procurement in the excluded sectors, COM(88) 376 final, 11 October 1988.
6.18 Council Directive 90/531/EEC, OJ L297/1 29 October 1990.
6.19 Proposal on the co-ordination of procedures on the award of public service contracts, COM(90) 372 final 6 December 1990.
6.20 Objection sustained, *New Civil Engineer*, 5 October 1989 pp.47–50.
6.21 Council Directive of 21 December 1989 on the co-ordination of the laws, regulations and administrative provisions relating to the application of review procedures to the award of public supply and public works contracts, OJ L395 30 December 1989 pp. 33–35.
6.22 COM(90) 287 OJ No C216, 31 August 1990.
6.23 *New Builder*, 8 February 1990, p.7.
6.24 as above.
6.25 EC public procurement rules may cause anxiety to buyers, *Financial Times*, 17 January 1991.
6.26 Commission of the European Communities, Public Procurement; Regional and Social Aspects, COM(89) 400 final, 24 July 1989.

Chapter 7

7.1 COM(89) 248 final 30 May 1989.
7.2 COM(89) 568 final, November 1989.
7.3 *Litster* v. *Forth Dry Dock Co Ltd* [1989]2 WLR, [1989] 1 All ER 1134.
7.4 Case 152/84 *Marshall* v. *Southampton and South West Hampshire Area Health Authority* [1986] ECR 723, [1986] 1 CMLR 688.
7.5 *Pickstone* v. *Freemans PLC* [1989] AC 66, [1988] 3 WLR 265.
7.6 Case C262/88 *Barber* v. *Guardian Royal Exchange Assurance Group* [1990] *The Times* law reports, 18 May 1990.
7.7 Commission of the European Communities, Social Europe Vol. 1 1990, Office for Official Publications of the European Communities, 1990 ISSN 0255-0776.
7.8 'EC attempt to block 12 hour night shifts', *New Builder* 2 August 1990, p.6.
7.9 Latest version of Proposal for a directive on procedures for informing and consulting employees ('Vredeling' directive) OJ C217 12.8.83 pp3–16.
7.10 OJ L183 29.6.89 page 1.
7.11 COM(90) 275 final, OJ C213 28.8.90 p. 2; modified proposal COM(91)117 final-SYN 279 OJ C112, 27 April 1991.
7.12 COM(89) 568 final, Brussels 29 November 1989.
7.13 Dr Chris Brewster and Dr Paul Teague, European Community Social Policy; Its impact on the UK, Institute of Personnel Management 1989.
7.14 as 7.13 above.

Chapter 8

8.1 Robert Read and Cathy Read, 'Breathing can be hazardous to your health', *New Scientist* 23 February 1991 pp 34–37.
8.2 Council Directive 85/337/EEC of 27 June on the assessment of the effects of certain public and private projects on the environment, OJ L175 5/7/85, p. 40.
8.3 OJ L158 23/6/90 p. 56.
8.4 OJ L120 11/5/90 p. 1.
8.5 OJ C251 4/10/89 p. 3.
8.6 OJ C001 4/1/90 p. 20.
8.7 Department of the Environment, *This Common Inheritance*, HMSO, 1990.
8.8 as 8.7 above.
8.9 John Gribbin, 'An assault on the climate consensus', *New Scientist*, 15 December 1990, pp. 26–31.

Chapter 9

9.1 Commission of the European Communities, Research on the 'cost of non-Europe', *Basic Findings Volume 1*, Office for Official Publications of the European Communities, 1988. ISBN 92 825 8605 7 (the bulk of the study was in English but the section on construction products was in French; however, the National Economic Development Office in London have produced an English translation).

9.2 *Cassis de Dijon* case: as reference 4.24.

9.3 Council resolution of 7 May 1985 OJ C136, 4 June 1985 p. 1.

9.4 Council Directive of 28 March 1983, 83/189/EEC, OJ L109 26/4/83, pp. 8–12 and amendments introduced by Council Directive of 22 March 1988, 88/182/EEC, OJ L81 26/3/88 p. 75.

9.5 Available from CEN, Rue Brederode 2, B-1000, Bruxelles, Belgium.

9.6 Council Directive of 21/12/88, OJ L40 11.2.89 p.12.

9.7 Commission of the European Communities, Commission Green Paper on the Development of European Standardisation: action for faster technological integration in Europe, COM(90)456 final, Brussels 8.10.90.

9.8 Council Decision of 13 December 1990, concerning the modules for the various phases of the conformity assessment procedures which are intended to be used in the harmonisation directives, OJ L380, 31 December 1990.

Chapter 10

10.1 *D & F Estates Ltd and Others* v. *The Church Commissioners for England and Others* [1988] 2 All ER 992.

10.2 *Greater Nottingham Co-op* v. *Cementation Ltd* [1988]2 All ER, p. 971.

10.3 *Murphy* v. *Brentwood District Council*, Construction Industry Law Letter, July 1990, p. 604.

10.4 85/374/EEC of 25 July 1985, OJ L210 7/8/85 pp.29–33.

10.5 *Rylands* v. *Fletcher* (1868) LR 3HL 330.

10.6 'Consumer Act poses product headaches, warn lawyers', *Building* 15 March 1988, p. 55.

10.7 Proposal for a directive on civil liability for damage caused by waste, OJ C251, 4 October 1989.

10.8 'Brussels must learn from US mistakes over waste damage', *Financial Times*, 10 January 1991.

10.9 Commission of the European Communities, Final Report III/8326/89-EN, 2 February 1990.

10.10 COM(90) 482 final – SYN 308, 20 December 1990.

Chapter 11

11.1 as 1.1 above.
11.2 85/384/EEC OJ L 223 21/8/85 p.15.
11.3 89/48/EEC OJ L 19 24/1/89 p. 16.
11.4 David Waller, 'Heading for the dawn of the Euro accountant', *Financial Times* 25 October 1990.

Chapter 12

12.1 Council Regulation (EEC) 2052/88 OJ L 185 15.7.88 p.9. Council Regulations (EEC) 4253/88, 4254/88, 4255/88, 4256/88 OJ L374 pages 1, 15, 21, 25. Copies of all these regulations and a guide to the reforms were published in the following: Commission of the European Communities, Guide to the reform of the Community's Structural Funds, Office for Official Publications of the European Communities, Luxembourg 1989, ISBN 92 826 0029 7.
12.2 World Aid Section, DTI, Room 042, 1-19 Victoria Street, London SW1H 0ET, tel: 071-215 4255/5615.
12.3 Chris Brewster and Paul Teague (1989) *European Community Social Policy*, Institute of Personnel Management, London 1989.
12.4 Mr N. Wegter, Tel: Brussels 235 5069, Fax: 235 0143.
12.5 Commission of the European Communities (DGXII), *EC research funding – a guide for applicants*, Commission of the European Communities, Brussels 1990.
12.6 Department of Trade and Industry, *Guide to European Community Research and Development Programmes*; available from Room 201, Ashdown House, 123 Victoria Street, London SW1E 6RB; Tel: 071-215 6618.
12.7 ECHO Customer Service, 177, route d'Esch, L-1023 Luxembourg, Tel. Luxembourg code + 352 48 80 41 (ECHO help desk), Fax + 352 48 80 40.

Index